# MORE <u>AWESOME</u> THAN <u>MONEY</u>

## ALSO BY JIM DWYER

*False Conviction*

*102 Minutes* (with Kevin Flynn)

*Actual Innocence* (with Peter Neufeld and Barry Scheck)

*Two Seconds Under the World* (with David Kocieniewski, Dee Murphy, and Peg Tyre)

*Subway Lives*

# MORE AWESOME THAN MONEY

Four Boys and Their Quest
to Save the World from Facebook

# Jim Dwyer

Viking

VIKING
Published by the Penguin Group
Penguin Group (USA) LLC
375 Hudson Street
New York, New York 10014

USA | Canada | UK | Ireland | Australia | New Zealand | India | South Africa | China
penguin.com
A Penguin Random House Company

First published by Viking Penguin, a member of Penguin Group (USA) LLC, 2014

Support for research on this book was provided by the Society of Professional Journalists through the Eugene G. Pulliam Fellowship.

LIBRARY OF CONGRESS CATALOGING-IN-PUBLICATION DATA
Dwyer, Jim, 1957–
More awesome than money : four boys and their heroic quest to save your privacy from Facebook / Jim Dwyer.
pages cm
Includes bibliographical references and index.
ISBN 978-0-670-02560-2 (hardback)
1. Diaspora (Project) 2. Internet industry—United States. 3. Online social networks—United States. 4. Privacy, Right of—United States. 5. New business enterprises—United States. 6. Business failures—United States. I. Title.
HD9696.8.U54D533 2014
384.3'8—dc23          2014004511

Printed in the United States of America
10  9  8  7  6  5  4  3  2  1

Set in Mercury Text G1 with DIN Next LT Pro
Designed by Daniel Lagin

*For Maura & Catherine*
*Blessed am I, amongst women.*

There's something deeper than making money off stuff.
Being a part of creating stuff for the universe is awesome.

—ILYA ZHITOMIRSKIY

# MORE AWESOME THAN MONEY

# INTRODUCTION

I f his laptop had been a mirror, the face staring back at Dan Grippi would have been some blend of boy and wild man. Nature had given him full, pouty, rebel-without-a-cause lips, and a jewelry shop in the East Village had given him a piercing in the bottom one a few days after he started college. He fastened a ring in the hole. His father was not thrilled.

Long and lean to begin with, Dan grew sideburns that ran down below each ear, sketch-strokes of whiskers that further drew his angular face to a point. He gave himself another inch of height by heaping his black hair in a pile, occasionally pinning it to the air with a generous, retro slathering of gel. On his ring finger he wore what appeared to be a cyanide capsule. Most of his arms rippled from short-sleeved white T-shirts. With the Elvis hair, the Springsteen sideburns, and the scary jewelry, Dan Grippi turned out for the world looking like a hybrid greaser-punk.

Except that he was neither. He smiled easily and warmly, spoke softly, but thought twice or three times before uttering a syllable. On the cross-country team, he logged miles in silence. His digital graphic art-work won awards for him while he was a teenager. A portfolio of that work had gotten him into college, overcoming indifferent grades at his high school in the Long Island suburbs of New York. His striking looks got him work as a model, and his mastery of the blend and stitch of music beats scored him gigs as a DJ. The menacing-looking cyanide ring was really a spare piece salvaged from a build-it-yourself printer he had helped assemble. He was a nerd with muscles.

On a February night in 2010, he stared at a page on Facebook, the soul-sorting social network machine. It was time for him to get out. He had joined in 2005, when he was sixteen. Facebook had started circling the globe a year earlier, college by college, working its way to high school students. It became the gyroscope of a generation, a tool for high-speed social navigation. Long after Dan's high school classmates had scattered to colleges all over the map, their friendship had a digital pulse on Facebook. At New York University in Greenwich Village, Dan would check out the Facebook listings of people who caught his eye in a class or at a party. He could see who they knew in common, sniff their electronic pheromones. By now, in his final months of college, he was so hooked on Facebook that even when he was entertaining guests in his apartment, he would often sit with a computer in his lap so he could keep track of what people were up to elsewhere. That was sick, he knew. But that wasn't the worst of it.

After a week of fiddling with the settings, clicking and unchecking boxes, disabling certain notifications, he kept coming to the same end point. It was hopeless. While he might be able to regulate some settings on his own account, he had no control over what his friends did with the applications that they ran on theirs. They could, for instance, permit a game to contact everyone in their address books, giving access to everything Dan had shared with friends to these third parties. Facebook was everywhere: go to a music website, and there would be a Facebook "like" button, meaning that it could follow him there. It knew what he read and knew what he listened to and knew what he watched. This was not just an extension of the high-octane party scene that defined his weekends. He could barely think a thought without giving some hint to the machinery that was recording it all. The site had grafted onto his personality.

Enough, he decided.

Using a piece of software, he scraped all the information and pictures he had posted on Facebook. That got him a copy of everything.

One last time, he checked updates from friends. Then he found his way to the account settings. Buried several layers down was the box he was looking for.

Delete account.

Was he sure? the software asked.

Dan clicked his affirmation.

Okay, then. The account would be deactivated but not actually deleted for two more weeks. Any activity by him during that time—logging in, checking on friends, the least flicker of digital life—would be interpreted by Facebook as a sign that he had changed his mind and didn't really want to shut it down. That was all boilerplate delivered to anyone who declared an intention to leave the flock.

The next message from Facebook was tailored for Daniel Grippi. "Your friends are going to miss you," it said.

Dan had about four hundred "friends" on Facebook, a number of them people he had never spoken with in the flesh, or, for that matter, online. Someone who knew someone would send a friend request. He'd say yes. And that would be the end of it. No loss.

But there were others: real buds from high school, people he cared about in college, family members. Authentic friends. Their pictures were the ones appearing on the screen. It was a simple matter for the Facebook servers to figure out whose profile he had checked out, with whom he had exchanged messages, even people who were in the same picture.

Facebook kept records of what he cared about. Knowing who and what Dan Grippi wanted in his life—knowing more than Dan himself consciously knew—was Facebook's business. It sold that knowledge.

Daniel, Facebook pleaded. Wouldn't you want to stay in touch with these people?

They were hostages; Facebook was transforming his attempt to quit the network into a betrayal of his fondness for them.

Manipulation.

He clicked the delete button.

In the winter of 2010, four talented young nerds with time on their hands decided that they should bring an end to Facebook's monopoly on social networking. They were man-boys, college students who became friends while staying up all night, eating pizza and hacking at geeky projects in a computer club at New York University. Their mission would have been literally inconceivable just one generation earlier; social networking barely existed when they began high school; Mark Zuckerberg had not yet started "The Facebook" in his dorm room at Harvard. But by 2010, Facebook had become such a global behemoth, joined by hundreds of people every minute, that a revolt seemed inevitable. Someone had to do it.

Like most "free" web services, Facebook was a middleman that made money by peeking into everybody's business. With the right software tools, the four guys believed, no one would need to surrender everything to big companies like Zuckerberg's. People could connect to friends and networks of friends without going through the servers of Facebook, Inc. It was a simple notion of vast disruption. Such an alternative would mean remapping the lines of power in digital society. To Dan Grippi and three other guys from the NYU computer club, Ilya Zhitomirskiy, Max Salzberg, and Raphael Sofaer, it would be a fun way to spend the summer of 2010. In the months before then, they spent nearly every conscious minute in room 311 of the math building at NYU working on their upstart social network project. They called it Diaspora.

"We're not about killing off Facebook," Raphael said. In Diaspora, they aimed to create a tool that would emancipate the power and vitality of social networks from the control of a single company. "Social networks have only existed for 10 years. We don't really know what's going to happen to our data, but it's going to exist into the foreseeable future. We need to take control of it."

His words were printed in the *New York Times,* and the notion that such a thing was possible thrilled hundreds of thousands of people. Social networking occupied an essential place in society, but was Facebook—avatar for the commercial surveillance embedded in the digital infrastructure of modern life—an irreplaceable link for all vital communication and connections? The Diaspora vision was thrillingly countercultural.

The four set out at a pivot point in human history, their project a single tiny pixel on a vast screen of rapid change, watched by millions. In June 2010, they moved to California to build the tools that would save the world from Facebook. Also, they hoped, to have fun. Dan and Max had just graduated. They had the most programming experience and chops. Dan, besides his strong eye for design, had a will of steel when it came to coding. Max, the eldest of the four, would pile thankless tasks on his back until he resembled an ant carrying five hundred times his weight. With a gift for turning phrases, Max described himself as an "effusive salesman type," and said that his rhetoric often benefited from a "cold bucket of water" dumped on it by Rafi, the youngest of the group, who possessed a cool detached intelligence. He was a quick study in programming. With some trepidation, he took a leave from school. So did Ilya, a math prodigy, Russian émigré, and idea-heaving furnace. Technically the weakest software programmer of the group, Ilya was their most potent spiritual force. No one could imagine Diaspora without him. Once they relocated to San Francisco, Ilya's home, an apartment on Treat Street in the Mission District, became a center of gravity for all manner of grand schemes and hopes. It was called the Hive.

The neighborhood of the Hive pulsed with people in their early twenties, constellations of dreamers and engineers and strivers in the expanding universe of the tech boom. Google ran a private bus system to bring employees to its offices forty miles south. But the area was not just home to the well-salaried and corporate-backed. Computers and computing had become dead cheap, a fraction of what they had been a decade earlier, and shrinking by the day. Brainpower, not money, was the essential capital.

Not only were there fortunes to be made; there were also fortunes to shape. Who could resist? Like steel at the start of the twentieth century, software had the power to win wars, bring prosperity, change lives. It could guess your appetite before the first pang of desire had struck.

This generation had come from cities around the world, out of schools and universities certainly, but also old factories and basements, derelict garages, and warehouses that had been turned into hacker spaces—rooms where they could tinker with software and cheap electronic parts, write programs and build robots or break into virtual cabinets where they didn't belong.

Said one hacker: these spaces are a force for chaos in the world; when you walk in, you expect to see kittens in jet packs flying around the room.

They slept in mansions tricked up as dorms, rented rooms in flophouse hotels, surfed from couch to couch, shared apartments at places like the Hive. Creature comforts were for later; right now, the act of creation, of making—not of making *it*—was all that mattered. Fall on your face? Get up. Fail faster. Like mentalists playing tricks with a spoon, the young people were sure they could bend the steel of their era just by thinking long and hard.

They had a suppleness of spirit unburdened by the thickness of middle-age expertise in what couldn't be done.

The Hive was a perfect space for parties, and Ilya was the ringmaster. Its spacious backyard included a tent-cabin, a fixture adopted from the utopianist desert festival Burning Man, and a necessity for outdoor gatherings in the foggy perma-chill of the Bay Area. Parties at the Hive customarily had a meme, or perhaps it was the other way around: someone would think of a meme and a party would follow. There was "We Are the Internet and We Come in Peace," and later, "FuckYeahCarlSagan." When someone in the circle revealed that his Wi-Fi password was "FuckTomCruise," the meme for the next party was declared "FuckYeahTomCruise," and Tom Cruise–themed drinks, like the Top Gun cocktail, were concocted, while the dome of the tent became the screen for a continuously looping video of the actor speaking about Scientology. On any given night, the sound system might be knocking out techno-jams, or music for swing or ballroom dance, or all of them.

Hardly any activity would not be considered: a woman who had cultivated an interest in a diet of insects was invited to whip up something on the barbecue, and she produced plates of grilled-cheese-and-mealworm-larvae sandwiches. These were washed down with craft beer and Two Buck Chuck, the value wine from Trader Joe's.

Raucous and humming, the Hive parties were descendants of the gatherings eighty years earlier of the young and poor, philosophers and scientists and artists, in a marine biologist's lab in Pacific Grove, forty miles to the south, caught in the amber of John Steinbeck's *Cannery Row*.

One weekend at the Hive, there was a contest to break the world record for links of plastic monkeys, each arm hooked into the next until the line stretched up three flights and right to the big bay window that looked out over the yard, the tent, and the city. The window opened from Ilya's bedroom. He had practically invented this place.

Convener of many parties, Ilya was the one who stayed up latest talking longest with roommates, or with strangers he'd met somewhere and had invited to hang out, or with people who'd wandered in.

One night, from behind the closed door to his bedroom at the Hive, he'd heard his lawyer roommate and math roommate talking in the living room. The lawyer asked the math guy about the importance of a high-level mathematical theorem; Ilya, dissatisfied with the explanation he heard through the door, burst out of the bedroom clad only in boxer shorts. He proceeded to sketch out the implications of the theorem on the floor. Then he returned to his bedroom, where he had been entertaining a young woman visitor. Of which there were many: he had little reticence about hitting on girls, and many found it hard to resist the package of a slightly disheveled guy who wore giant plastic orange sunglasses and aquamarine neon pants, a brainy polymath who actually wanted to learn what other people were passionate about.

"Gather epic people," he often declared, "and make unreasonable demands." He was a nerd who glowed with warmth and humor. The world he and the other Diaspora guys had settled into was one of pure buoyancy; the work they were doing was all gravity.

In the tenth century, Rabbeinu Gershom, a sage in Mayence, Germany, set out a series of prohibitions that would be enduring parts of Jewish life. They were issued in response to human impulses that he deemed to be socially destructive, not to mention an offense against the Creator. For instance, Rabbeinu Gershom barred polygamy and forbade divorce of a woman without her consent.

The rabbi also banned the reading of other people's private mail.

A millennium after Rabbeinu Gershom, other people's business was sheathed in fiber-optic cable. Ancient human urges to snoop had lost none of their voltage, but the prohibitions and social inhibitions were dissolving. In a virtual instant, forward-thinking businesses data-mined, data-scraped, data-aggregated. As these became exalted missions, digital culture erased social and legal boundaries that had been honored, however imperfectly, for centuries. Commercial surveillance was built into the ecology of everyday life.

Like nothing else since humans first stood upright, the World Wide

Web has allowed people to connect and learn and speak out. Its dominant architecture has also made it possible for businesses and governments to fill giant data vaults with the ore of human existence—the appetites, interests, longings, hopes, vanities, and histories of people surfing the Internet, often unaware that every one of their clicks is being logged, bundled, sorted, and sold to marketers. Together, they amount to nothing less than a full-psyche scan, unobstructed by law or social mores.

"Facebook holds and controls more data about the daily lives and social interactions of half a billion people than 20th-century totalitarian governments ever managed to collect about the people they surveilled," Eben Moglen, a technologist, historian, and professor of law at Columbia University, observed in 2010.

That shirtless, drunken picture? The angry note to a lover, the careless words about a friend, the so-called private messages? Facebook has them all. And its marketers know when it is time to pop an advertisement for a florist who can deliver make-up flowers, or for a private investigator.

Students at MIT figured out the identities of closeted gay people by studying what seemed like neutral information disclosed on Facebook. At the headquarters of Facebook in Palo Alto, Mark Zuckerberg was said to amuse himself with predictions on which Facebook "friends" would eventually hook up by watching how often someone was checking a friend's profile, and the kinds of messages they were sending.

The uproar over Facebook's privacy policies obscured intrusions on an even grander scale by other powerful forces in society. Everyone had heard of AOL and Microsoft; few were familiar with their subsidiaries. Atlas Solutions, purchased by Microsoft in 2007, told businesses that it deploys "tracking pixels"—a kind of spy cookie that cannot be detected by most filters—to follow Internet users as they look at billions of web pages per month. These invisible bugs watch as we move across the web, shopping, reading, writing. Our habits are recorded. The pixels live in computer caches for months, waiting to be pulsed. Facebook bought Atlas in 2013, helping it track users when they left the site.

And virtually unknown to users, AOL's biggest business was never the cheery voice announcing, "You've got mail"; it was the billions of data items its subsidiary, Platform A, mined from Internet users, linking their interests and purchases, zip codes and significant others. The data was stored on servers physically located in giant warehouses near Manassas, Virginia. AOL boasted that it followed "consumer behavior across thousands of websites."

Facebook was a proxy in a still larger struggle for control over what used to be the marrow of human identity: what we reveal and what we conceal, what we read and what we want. Just as human tissue is inhabited by trillions of bacteria, so, too, our online life is heavily colonized by external forces, invisible bits of code that silently log our desires and interests, and, at times, manipulate them.

An experiment conducted in 2010 by the *Wall Street Journal* showed how far commercial interests could penetrate personal information, unbeknownst to web users. As part of a remarkable series called "What They Know," the *Journal* team set up a "clean" computer with a browser that had not previously been used for surfing. The results: after visiting the fifty most popular websites in the United States, the reporters found that 131 different advertising companies had planted 2,224 tiny files on the computer. These files, essentially invisible, kept track of sites the users visited. This allowed the companies to develop profiles that included age, gender, race, zip code, income, marital status, health worries, purchases, favorite TV shows, and movies. Deleting the tracking files did not eliminate them: some of them simply respawned.

Handling all the data they collected was possible because computing power continued to double every eighteen months to two years, the rate predicted in 1965 by the technologist Gordon Moore. Cheap and prolific by 2010, that power enabled the creation of bare-bones start-ups and the granular monitoring of personal habits. "We can uniquely see the full and complete path that your customers take to conversion, from seeing an ad on TV to searching on their smartphone to clicking on a display ad on their laptop," a business called Converto boasted on its website in 2013.

The online advertising industry argued that the ability to tailor ads that appeared on a screen to the presumed appetites of the person using the computer was the foundation of the free Internet: the advertising dollars supported sites that otherwise would have no sources of revenue.

Whatever the merits of that argument, it was hard to defend the stealthiness of the commercial surveillance. No national law requires that this monitoring be disclosed, much less forbids it. A few halfhearted efforts by the Federal Trade Commission to regulate the monitoring have gone nowhere. There was no way for people to get their data back.

Or even their thoughts.

In mid-2013, two researchers published a paper entitled "Self Censorship on Facebook," reporting that in a study of 3.9 million users, 71 percent did not post something that they started to write. That is, they changed their minds. While this might look like prudence, or discretion, or editing, the researchers—both working at Facebook—described it as "self-censorship" and wrote that such behavior was a matter of concern to social network sites. When this happens, they wrote, Facebook "loses value from the lack of content generation."

The company maintained that users are told that it collects not only information that is openly shared but also when you "view or otherwise interact with things." That means, the company asserted, the right to collect the unpublished content itself. "Facebook considers your thoughtful discretion about what to post as bad, because it withholds value from Facebook and from other users. Facebook monitors those unposted thoughts to better understand them, in order to build a system that minimizes this deliberate behavior," Jennifer Golbeck, the director of the Human-Computer Interaction Lab at the University of Maryland, wrote in *Slate*.

At every instant, the fluid dynamics of the web—the interactions, the observations, the predations—are logged by servers. That such repositories of the lightning streams of human consciousness existed was scarcely known and little understood. "Almost everyone on Planet Earth has never read a web server log," Eben Moglen, the Columbia law professor said. "This is a great failing in our social education about technology. It's equivalent to not showing children what happens if cars collide and people don't wear seat belts."

One day in the 1970s, a man named Douglas Engelbart was walking along a beachfront boardwalk in California when he spotted a group of skateboarders doing tricks.

"You see these kids skateboarding actually jumping into the air and the skateboard stays attached to their feet, and they spin around, land on the skateboard, and keep going," Engelbart remembered many years later.

For Engelbart, those skateboarders were a way to understand the unpredictability of technology. "I made them stop and do it for me six times, so I could see how they did it. It's very complicated—shifting weight on their feet, and so on. You couldn't give them the engineering and tell them to go out and do that. Fifteen years ago, who could have designed that? And that's all we can say about computers."

A little-celebrated figure in modern history, Engelbart had spent decades thinking about how computers could be linked together to augment human intelligence, to solve problems of increasing complexity. At a gathering of technologists in 1968, he showed what could happen when computers talked to one another. For the occasion, he demonstrated a device that made it easier for the humans to interact with the computer. The device was called the mouse. The cofounder of Apple, Steve Wozniak, said that Engelbart should be credited "for everything we have in the way computers work today."

The emergence of the personal computer and the Internet, with its vast democratizing power, were part of Engelbart's vision. He died in July 2013, a few weeks after revelations by a man named Edward Snowden that the United States National Security Agency was collecting spectacular amounts of data. Writing in the *Atlantic,* Alexis C. Madrigal noted: "We find ourselves a part of a 'war on terror' that is being perpetually, secretly fought across the very network that Engelbart sought to build. Every interaction we have with an Internet service generates a 'business record' that can be seized by the NSA through a secretive process that does not require a warrant or an adversarial legal proceeding."

The business purposes of such data collection are apparent, if unsettling. But what need did governments have for it? Among Western democracies, the stated purpose was piecing together suggestive patterns that might reveal extremists plotting attacks like those carried out on September 11, 2001. The dystopic possibilities of such powers had, of course, been anticipated by George Orwell in *1984,* and by the visionary cyberpunk novelist William Gibson in *Neuromancer.* But fiction was not necessary to see what could be done: in 2010, its utility as an instrument of surveillance and suppression had been realized in, among other places, Syria, Tunisia, Iran, and China.

So, too, were its other properties: as the Diaspora guys were making plans for their project in 2010, the Arab Spring was stirring to life, some of it in subversive online communications that either were not noticed or not taken seriously by the regimes that would soon be toppled. The same mechanisms allowed more alert regimes to surveil opponents, or to be led directly to the hiding places of dissidents who had unwittingly emitted location beams from phones in their pockets.

By 2010, in just the two years since Raphael Sofaer had entered college, Facebook had grown by 300 million users, almost five new accounts every second of the day. The ravenous hunger for new ways to connect in a sprawling world was not invented by Facebook, but the company was perfectly positioned to meet it, thanks to skill, luck, and the iron will of its young founder, Mark Zuckerberg. A manager in Facebook's growth department, Andy Johns, described going to lunch for the first time with his new boss, one of Zuckerberg's lieutenants.

"I remember asking him, 'So what kind of users am I going after? Any particular demographics or regions? Does it matter?' and he sternly responded 'It's fucking land-grab time, so get all of the fucking land you can get.'"

Could four young would-be hackers build an alternative that preserved the rich layers of connection in social networking without collecting the tolls assessed by Facebook? Would anyone support their cause?

When word got out about their project, they were swamped.

In a matter of days, they received donations from thousands of people in eighteen countries; tens of thousands more started to follow their progress on Twitter, and in time, a half million people signed up to get an invitation. That was more weight than the four were ready to carry. On the night that their fund-raising drive exploded, as money was pouring in through online pledges, nineteen-year-old Rafi Sofaer toppled off the even keel where he seemed to live his life. It was too much. They were just trying to build some software. "Make them turn it off!" he implored the others. It couldn't be done.

Four guys hanging around a little club room at NYU suddenly found themselves handed a global commission to rebottle the genie of personal privacy. They were the faces of a movement, a revolution against the settled digital order. Their job was to demonetize the soul.

The problem they set out to solve was hard. That was its attraction. They were young, smart, quick to learn what they did not know, and girded for battle. Suddenly, they had a legion of allies. And expectations. It was delightful, for a while.

# PART ONE

## To the Barricades

# CHAPTER ONE

Sharply turned out in a tailored charcoal suit accented with a wine-red tie, the burly man giving the lecture had enchanted for twenty minutes, one moment summoning John Milton from the literary clouds, the next alluding to the lost continent of Oceania, then wrapping in Bill Gates and Microsoft. Every offhand reference was, in fact, a purposeful stitch in a case for how the entire architecture of the Internet had been warped into a machine for surveilling humans doing what they do: connecting, inquiring, amusing themselves. Then he made the subject concrete.

"It is here," the speaker said, "of course, that Mr. Zuckerberg enters."

Seated in an NYU lecture hall in Greenwich Village on a Friday evening, the audience stirred. Most of those attending were not students but members of the Internet Society, the sponsor of the talk. But no one listened more avidly than two NYU students who were seated near the front, Max Salzberg and Ilya Zhitomirskiy.

And they were keen to hear more about "Mr. Zuckerberg." That, of course, was Mark Zuckerberg, boy billionaire, founder and emperor of Facebook, and a figure already well known to everyone in the crowd that had come to hear a talk by Eben Moglen, a law professor, history scholar, and technologist. *The Social Network*, a fictionalized feature film about the creation of Facebook, was still eight months away from its premier. Nevertheless, the name Zuckerberg needed no annotation. And at age twenty-five, he had never gotten an introduction of the sort that Moglen was about to deliver to him in absentia.

"The human race has susceptibility to harm, but Mr. Zuckerberg has attained an unenviable record," Moglen said. "He has done more harm to the human race than anybody else his age. Because—"

Moglen's talk was being live-streamed, and in an East Village apartment a few blocks away, an NYU senior named Dan Grippi, who had been only half listening, stopped his homework.

"Because," Moglen continued, "he harnessed Friday night.

"That is, everybody needs to get laid, and he turned it into a structure for degenerating the integrity of human personality."

Gasps. A wave of laughs. A moment earlier, this had been a sober, if engaging, talk, based on a rigorous analysis of how freedom on the Internet had been trimmed until it bled. As lawyer, hacker, and historian, Moglen possessed a rare combination of visions. He blended an engineer's understanding of the underlying, intricate architecture of the Internet and the evolving web with a historian's panoramic view of how those structures supported, or undermined, democratic principles and a lawyer's grasp of how far the law lagged behind technology. For nearly two decades, Moglen had been the leading consigliere of the free-software movement in the United States, and even if not everyone in the auditorium at New York University was personally acquainted with him, they all knew of him. A master orator, Moglen knew that he had just jolted his audience.

He immediately tacked toward his original thesis, this time bringing along Zuckerberg and Facebook as Exhibit A, saying: "He has, to a remarkable extent, succeeded with a very poor deal."

Most of the regalia of Facebook, its profile pages and activity streams and so on, were conjured from a commonplace computer language called PHP, which had been created when Mark Zuckerberg was eleven years old. By the time he built the first Facebook in 2004, PHP was already in use on more than 10 million websites; much of the web came to billowing life on computer screens thanks to those same text scripts, always tucked out of sight behind Oz's curtain. Knowing that, Moglen put the terms of the Zuckerberg/Facebook deal with the public in currency that his audience, many of them technologists, could grasp in an instant.

"'I will give you free web hosting and some PHP doodads, and you get spying for free, all the time.' And it works."

The audience howled.

"That's the sad part, it works," Moglen said. "How could that have happened? There was no architectural reason, really."

As lightning bolts go, this one covered a lot of ground. A mile away in his apartment, Dan listened and thought, what if he's right? The guy who created PHP called it that because he needed some code scripts for his personal home page. Which is sort of what Facebook was: You had a home page that could be played with in certain, limited ways. Post a picture. Comment on a friend's. Read an article that someone liked or hated. Watch a funny cat video. But all these things were possible on any web page, not just Facebook, which was really just a bunch of web pages that were connected to one another. Maybe there was no good technical reason that social networks should be set up the way Facebook was. For Moglen, it was an example—just one, but a globally familiar one—of what had gone wrong with the Internet, a precise instance of what he had been talking about for the first twenty minutes of his speech, in every sentence, even when he seemed to be wandering.

Dan was starting to wish he had gone to the talk in person. He knew that Max and Ilya were there. Practically from the moment Moglen had opened his mouth to make what sounded like throw-away opening remarks, they had been galvanized. "I would love to think that the reason that we're all here on a Friday night is that my speeches are so good," Moglen had said. The audience tittered. In truth, the speeches of this man were widely known not just as good but as flat-out brilliant, seemingly unscripted skeins of history, philosophy, technology, and renaissance rabble-rousing, a voice preaching that one way lay a dystopic digital abyss, but that just over there, in striking distance, was a decent enough utopia.

"I actually have no idea why we're all here on a Friday night," Moglen continued, "but I'm very grateful for the invitation. I am the person who had no date tonight—so it was particularly convenient that I was invited for now. Everybody knows that. My calendar's on the web." No need for Moglen to check any other calendars to know that quite a few members of the audience did not have dates, either. His confession was an act of kinship, but it also had a serious edge.

"Our location is on the web," Moglen said. Cell phones could pinpoint someone's whereabouts. Millions of times a year, the major mobile phone companies asked for and were given the precise location of people with telephones. There was no court order, no oversight, just people with law enforcement ID cards in their pockets.

"Just like that," he said, getting warmed up.

He was making these points three years before Edward Snowden emerged from the shadows of the National Security Agency to fill in the shapes that Mogen was sketching.

"The deal that you get with the traditional service called 'telephony' contains a thing you didn't know, like spying. That's not a service to you but it's a service and you get it for free with your service contract for telephony."

For those who hacked and built in garages or equivalent spaces, Moglen was an unelected, unappointed attorney general, the enforcer of a legal regimen that protected the power of people to adjust the arithmetic that made their machines work.

As the volunteer general counsel to the Free Software Foundation, Moglen was the legal steward for GNU/Linux, an operating system that had been largely built by people who wrote their own code to run their machines. Why pay Bill Gates or Steve Jobs just so you could turn your computer on? For the low, low price of zero, free software could do the trick just as well, and in the view of many, much better. And GNU/Linux was the principal free system, built collaboratively by individuals beginning in the mid-1980s. It began as GNU, a code bank overseen by a driven ascetic, Richard A. Stallman, and found a path into modern civilization when a twenty-one-year-old Finnish computer science student, Linus Torvalds, adopted much of the Stallman code and added a key piece known as the kernel, to create a free operating system. (One of his collaborators called Torvalds's contribution Linux, and as the GNU/Linux release became the most widespread of the versions was routinely shorthanded as Linux, to the dismay of Stallman.) They were joined by legions of big businesses and governments following the hackers down the free-software road. On average, more than nine thousand new lines of code were contributed to Linux every day in 2010, by hundreds of volunteers and by programmers working for businesses like Nokia, Intel, IBM, Novell, and Texas Instruments.

The successor to Bill Gates as CEO of Microsoft, Steve Ballmer, fumed that Linux had, "you know, the characteristics of communism that people love so very, very much about it. That is, it's free."

It was indeed. As the Free Software Foundation saw things, in principle, the strings of 1s and 0s that make things happen on machines were no more the property of anyone than the sequence of nucleotides that provide the instructions for life in deoxyribonucleic acid, DNA.

Linux was the digital genome for billions of phones, printers, cameras, MP3 players, and televisions. It ran the computers that underpinned Google's empire, was essential to operations at the Pentagon and the New York Stock Exchange, and served as the dominant operating system for computers in Brazil, India, and China. It was in most of the world's supercomputers, and in a large share of the servers. In late 2010, Vladimir Putin ordered that all Russian government agencies stop using Microsoft products and convert their computers to Linux systems by 2015.

Linux had no human face, no alpha dog to bark at the wind; it had no profit-and-loss margins, no stock to track in the exchanges, and thus had no entries on the scorecards kept in the business news sections of the media. It was a phenomenon with few precedents in the modern market economy, a project on which fierce competitors worked together. In using GNU/Linux, they all had to agree to its licensing terms, whose core principles were devised primarily by Stallman, of the Free Software Foundation, in consultation with Moglen and the community of developers.

The word "free" in the term "free software" often threw people off. It referred not to the price but to the ability of users to shape the code, to remake, revise, and pass it along, without the customary copyright limitations of proprietary systems. Think of free software, Stallman often said, not as free as in free beer, but free as in free speech. So the principles of free software were spelled out under the license that people agreed to when they used it: anyone could see it, change it, even sell it, but they could not make it impossible for others to see what they had done, or to make their own subsequent changes. Every incarnation had to be available for anyone else to tinker with. Ballmer of Microsoft called it "a cancer that attaches itself in an intellectual property sense to everything it touches."

As the chief legal engineer for the movement, who helped to enforce the license and then to revise it, Moglen was the governor of a territory that was meant to be distinctly ungovernable, or at least uncontrollable, by any individual or business.

Having started as a lawyer for the scruffy, Moglen often found himself, as the years went by, in alliances that included powerful corporations and governments that were very pleased to run machines with software that did not come from the laboratories of Microsoft in Redmond, Washington, or of Apple in Cupertino, California. It was not that Moglen or his original long-haired clients had changed or compromised their views: the world simply had moved in their direction, attracted not necessarily by the soaring principles of "free as in free speech," or even because it was "free as in free beer." They liked it because it worked. And, yes, also because it was free.

The hackers had led an unarmed, unfunded revolution: to reap its rewards, all that the businesses—and anyone else—had to do was promise to share it. The success of that movement had changed the modern world.

It also filled the lecture hall on a Friday night. Yet Moglen, as he stood in the auditorium that night in February 2010, would not declare victory. It turned out that not only did free software not mean free beer, it didn't necessarily mean freedom, either. In his work, Moglen had shifted his attention to what he saw as the burgeoning threats to the ability of individuals to communicate vigorously and, if they chose, privately.

"I can hardly begin by saying that we won," Moglen said, "given that spying comes free with everything now. But we haven't lost. We've just really bamboozled ourselves and we're going to have to unbamboozle ourselves really quickly or we're going to bamboozle other innocent people who didn't know that we were throwing away their privacy for them forever."

His subject was freedom not in computer software but in digital architecture. Taken one step at a time, his argument was not hard to follow.

In the early 1960s, far-flung computers at universities and government research facilities began communicating with one another, a network of peers. No central brain handled all the traffic. Every year, more universities, government agencies, and institutions with the heavy-duty hardware joined the network. A network of networks grew; it would be called the Internet.

The notion that these linked computers could form a vast, open library, pulsing with life from every place on earth, gripped some of the Internet's earliest engineers. That became possible in 1989, when Tim Berners-Lee developed a system of indexing and links, allowing computer users to discover what was available elsewhere on the network. He called it the World Wide Web. By the time the public discovered the web in the mid-1990s, the personal computers that ordinary people used were not full-fledged members of the network; instead, they were adjuncts, or clients, of more centralized computers called servers.

"The software that came to occupy the network was built around a very clear idea that had nothing to do with peers. It was called server-client architecture," Moglen said.

So for entry to the promise and spoils of the Internet, individual users had to route their inquiries and communications through these central servers. As the servers became more powerful, the equipment on the desktop became less and less important. The servers could provide functions that once had been built into personal computers, like word processing and spreadsheets. With each passing day, the autonomy of the users shrunk. They were fully dependent on central servers.

"The idea that the network was a network of peers was hard to perceive after a while, particularly if you were a, let us say, an ordinary human being," Moglen said. "That is, not a computer engineer, scientist, or researcher. Not a hacker, not a geek. If you were an ordinary human being, it was hard to perceive that the underlying architecture of the net was meant to be peerage."

Then, he said, the problem became alarming, beginning with an innocent, and logical, decision made by naïve technologists. They created logs to track the traffic in and out of the servers. "It helps with debugging, makes efficiencies attainable, makes it possible to study the actual operations of computers in the real world," Moglen said. "It's a very good idea."

However, the logs had a second effect: they became a history of every inquiry that users made, any communications they had—their clicks on websites to get the news, gossip, academic papers; to buy music or to stream pornography; to sign in to a bird-watchers' website or to look at the latest snapshots of the birth of the universe from NASA and of the outfit that Lady Gaga wore to a nightclub the night before. The existence of these logs was scarcely known to the public.

"We kept the logs, that is, information about the flow of information on the net, in centralized places far from the human beings who controlled, or thought they controlled, the operation of the computers that increasingly dominated their lives," Moglen said. "This was a recipe for disaster."

No one making decisions about the architecture of the Internet, Moglen said, discussed its social consequences; the scientists involved were not interested in sociology or social psychology, or, for the most part, freedom. "So we got an architecture which was very subject to misuse. Indeed, it was in a way begging to be misused, and now we are getting the misuse that we set up."

The logs created as diagnostic tools for broken computers were quickly transformed into a kind of CT scan of the people using them, finely scaled maps of their minds. "Advertising in the twentieth century was a random activity; you threw things out and hoped they worked. Advertising in the twenty-first century is an exquisitely precise activity."

These developments, Moglen said, were not frightening. But, he warned: "We don't remain in the innocent part of the story for a variety of reasons.

"I won't be tedious and Marxist on a Friday night and say it's because the bourgeoisie is constantly engaged in destructively reinventing and improving its own activities. And I won't be moralistic on a Friday night and say that it is because sin is ineradicable and human beings are fallen creatures and greed is one of the sins we cannot avoid committing. I will just say that as an ordinary social process, we don't stop at innocent. We go on. Which is surely the thing you should say on a Friday night. And so we went on.

"Now, where we went on—is really toward the discovery that all of this would be even better if you had all the logs of everything. Because once you have the logs of everything, then every simple service is suddenly a gold mine waiting to happen. And we blew it, because the architecture of the net put the logs in the wrong place. They put the logs where innocence would be tempted. They put the logs where the fallen state of human beings implies eventually bad trouble. And we got it."

The locus of temptation, to dawdle with Moglen in the metaphysical, is not an actual place: the servers that held all this succulent data were not necessarily in a single physical location. Once the data was dragnetted from someone's Facebook entries, for instance, they could be atomized, the pieces spread across many servers, and then restored in a wink by software magic. The data was in a virtual place, if one that was decidedly not virtuous. The data was in the cloud, and thus beyond the law.

"You can make a rule about logs, or data flow, or preservation, or control, or access, or disclosure," Moglen said, "but your laws are human laws, and they occupy particular territory and the server is in the cloud and that means the server is always one step ahead of any rule you make or two, or three, or six, or poof! I just realized I'm subject to regulation, I think I'll move to Oceania now.

"Which means that, in effect, we lost the ability to use either legal regulation or anything about the physical architecture of the network to interfere with the process of falling away from innocence that was now inevitable."

In 1973, at age fourteen, Moglen had gotten a job writing computer programs for the Scientific Timesharing Corporation in Westchester, north of New York City, work that he continued for one company and another for the next decade. By 1986, at age twenty-six, he was a young lawyer, clerking for the Supreme Court justice Thurgood Marshall, and also working his way toward a PhD in history, with distinction, from Yale. His dissertation was titled "Settling the Law: Legal Development in New York, 1664–1776." At midlife, his geeky side, his legal interests, his curiosity about how human history was shaped, had brought him to the conclusion that software was a root activity of humankind in the twenty-first century, just as the production of steel had been an organizing force in the twentieth century. Software would undergird global societies.

Software, then, was not simply a rattle toy for playpens filled with geeks, the skeleton of amusements for a naïve public, but a basic moral and economic force whose complexity had to be faced coolly, with respect, not fear.

In the eighteenth century, Jeremy Bentham, a British social theorist, conceived of a prison where all the inmates could be seen at once, but without knowing that they were being observed. He called it the panopticon and predicted it would be "a new mode of obtaining power of mind over mind, in a quantity hitherto without example." Beginning in the 1970s onward, the cypherpunks, many of them pioneers at leading technology companies, saw that dystopian possibilities were built into the treasures of a networked world.

Moglen said: "Facebook is the web with 'I keep all the logs, how do you feel about that?' It's a terrarium for what it feels like to live in a panopticon built out of web parts.

"And it shouldn't be allowed. It comes to that. It shouldn't be allowed. That's a very poor way to deliver those services. They are grossly overpriced at 'spying all the time.' They are not technically innovative. They depend upon an architecture subject to misuse and the business model that supports them is misuse. There isn't any other business model for them. This is bad.

"I'm not suggesting it should be illegal. It should be obsolete. We're technologists, we should *fix* it."

The crowd roared. Moglen said he was glad they were with him, but he hoped they would stay with him when he talked about how to fix it. "Because then," he said, "we could get it done."

By now, Dan in his apartment, Ilya and Max in the auditorium, were mesmerized. His own students, Moglen said, comforted themselves that even though their Gmail was read by Google software robots for the purpose of inserting ads that were theoretically relevant to the content of their e-mails, no actual humans at Google were reading their correspondence. No one could entertain such a delusion about Facebook. News accounts based on various internal documents and sources suggested a streak of voyeurism on the premises.

"Facebook workers know who's about to have a love affair before the people because they can see X obsessively checking the Facebook page of Y," Moglen said. Any inferences that could be drawn, would be.

Students "still think of privacy as 'the one secret I don't want revealed,' and that's not the problem. Their problem is all the stuff that's the cruft, the data dandruff, of life, that they don't think of as secret in any way, but which aggregates to stuff that they don't want anybody to know," Moglen said. Flecks of information were being used to create predictive models about them. It was simple to deanonymize data that was thought to be anonymized, and to create maps of their lives.

The free-software movement could be proud of the tools it had created and protected from being absconded. It was not enough. "We have to fess up: if we're the people who care about freedom, it's late in the game, and we're behind," Moglen said. "I'm glad the tools are around but we do have to admit that we have not used them to protect freedom because freedom is decaying."

An illusion of convenience had eroded freedom, he said. "Convenience is said to dictate that you need free web hosting and PHP doodads in return for spying all the time because web servers are so terrible to run. Who could run a web server of his own and keep the logs? It would be brutal!" The crowd laughed: so many there actually did run web servers.

"What do we need?" Moglen asked.

"We need a really good web server you can put in your pocket and plug in anyplace. It shouldn't be any larger than the charger for your cell phone and you should be able to plug it into any power jack in the world, and any wire near it, or sync it up with any Wi-Fi router that happens to be in its neighborhood." Inside the little box would be software that would turn itself on, would collect stuff from social networks, and would send a backup copy of vital stuff—encrypted—to a friend's little box.

It all might have sounded far-fetched, except that the plug-in computers were already being made; they cost ninety-nine dollars, a price that was sure to drop, and needed only the right collection of free software to run them. He ran through the requirements: a program for social networks, for blogging, streaming music, and so forth. The servers of the world were already running on the free software of GNU/Linux. "The bad architecture is enabled, powered by us," Moglen noted. "The re-architecture is, too. If we have one copy of what I'm talking about, we'd have all the copies we need. We have no manufacturing or transport or logistics constraints. If we do the job, it's done. We scale."

That is: one copy of a piece of free software, and everything afterward is distributed over the air.

"It's a frontier for technical people to explore. There is enormous social payoff for exploring. The payoff is plain because the harm being ameliorated is current and people you know are suffering from it."

He reflected for a moment on the history of the free-software movement in meeting such challenges, and then moved back to the case in point. "Mr. Zuckerberg richly deserves bankruptcy," and the crowd applauded.

"Let's give it to him."

A voice shouted from aisles: "For free!"

"For free," Moglen agreed.

This effort was not about Facebook. The architecture of the web provided scaffolding for "immense cognitive auxiliaries for the state—enormous engines of listening for governments around the world. The software inside the plug-in computer could include special routing devices that disguise the digital traffic, making it harder to trace any individual on the Internet. "By the time you get done with all of that, we have a freedom box. We have a box that actually puts a ladder up for people who are deeper in the hole than we are."

All this from free software. "The solution is made of our parts. We've got to do it. That's my message. It's Friday night. Some people don't want to go right back to coding, I'm sure. We could put it off until Tuesday, but how long do you really want to wait? You know every day that goes by, there's data we'll never get back."

The first critical problem was identifying a way to attack it. "The direction in which to go is toward freedom—using free software to make social justice."

Someone shouted, "Yeah," and the applause washed across the room.

"But you know this," Moglen said. "That's the problem with talking on a Friday night. You talk for an hour and all you tell people is what they know already."

As the applause petered out, Max and Ilya felt like gongs that had been struck. They had arrived early for the talk, prodded by two teachers important to them: the adviser to the campus computer club, Evan Korth, who was also an officer of the New York division of the Internet Society; and Biella Coleman, an anthropologist who studied hacker culture and was Max's senior paper adviser. Now they did not budge. The moment, the possibility, the necessity of what Moglen had mapped out was nothing less than an alternative universe. It called to their idealism, and held the transgressive promise of, maybe, subverting a powerful institution. Its gravity absorbed them. So, yes, Friday would be spent in the math and computer science building. They had plenty to explore right there and then.

"Max," someone said from behind them.

Fred Benenson, who had graduated from NYU a few years earlier, had met Max at meetings of the campus branch of Students for Free Culture, a movement to ease copyright restrictions on the use of creative material.

Benenson could see that Max had been roused by Moglen's talk, and he could not resist playing devil's advocate. It was fine to talk about anonymity and privacy, he said, but in the real world, online retailers like Amazon and Netflix collected data from their customers, and used it to make recommendations on books, music, and movies.

"It's incredible how powerful those recommendation engines are," Benenson said. He had recently started at a job where he did research on just this kind of information gathering.

"You need data in aggregate form, even if it's anonymized, to make these interesting features that the users expect," he said.

"I know," Max said.

Yet they both knew that even if the records were kept anonymously, it was possible to match them up with other data, and identify people who had not realized how vulnerable they were to being easily deanonymized through reverse engineering, often by comparing anonymous and public databases. One famous example involved movies people watched and rated on Netflix. In 2007, two researchers at the University of Texas, Austin, showed they could easily figure out a person's supposedly private Netflix movie-viewing history by using other publicly available data.

A year earlier, Netflix had released 100 million movie ratings, from 500,000 people, and announced a $1 million prize for anyone who could improve the formula for making recommendations. The customer identities were scrubbed from the rankings. But the researchers were able to unmask them by comparing the Netflix movies they had privately rated with ones they had publicly discussed on IMDb, a website that catalogs and ranks movies.

That study was not the only example of "anonymous" data being laced with other, public information to trace its origins. AOL had released 20 million search queries after first stripping away identifying information like the users' names and their computer addresses. They were then assigned random numbers. It did not take long for Michael Barbaro, a business reporter with the *New York Times,* to track down the user who had searched for "numb fingers" and for "60 single men" and "dog that urinates on everything." The same person also queried women's underwear, landscapers in Lilburn, Georgia, and homes in a certain subdivision. The digital bread crumbs led directly to a sixty-two-year-old woman living with three dogs.

People wanted the convenience of recommendations, and many also wanted their privacy, and even sincere promises by web companies that data was being used only in big, random, atomized clumps was no protection, since information that had been teased apart could be reconstituted. The advances in computing power meant that a jar of bread crumbs could be turned into a full loaf.

"How are you going to reconcile that aggregation of data with the desire to remain private?" Benenson said.

Max was ready. People should be allowed to choose if they wanted to be included in the mass repositories, he said. That is, they ought to be asked what they wanted.

"It should be a conscious, explicit, opt-in," Max said.

"That's a great answer," Benenson said, surprised. "You've really thought about this. Wow. I'm inspired."

After the lecture hall emptied, the computer club convened in the little office on the third floor of the computer science building, electricity surging. High spirits reigned. Rafi and his brother Mike had come to NYU after a splendid meal with their father.

"I just got there for the end of his speech," Rafi said.

"It was awesome," Ilya said.

By temperament, Rafi was measured, even after a few glasses of wine at dinner, a cool yin to Ilya's hot yang. For that matter, he had developed immunities to lawyerly eloquence: his father, Abe Sofaer, was a retired federal judge, former senior counsel in the U.S. State Department, and partner in a big Washington law firm. Rafi, the youngest of five sons, also didn't need to hear the others rave about a talk that he had mostly missed.

"No speech could be better than the meal I just had," Rafi said.

Still, the mood was contagious. Even he was intrigued. They all were. And they kept talking about it, late into the night. It was February 2010. The four young men most moved by Moglen's speech had met a few months earlier in the computer club office, a campus chapter of the ACM, or Association for Computing Machinery.

# CHAPTER TWO

J ust as in cartoons when toys come to secret, robust life after dark, being inside the Courant Institute for Mathematics in the middle of the night had an emancipating effect on members of the computer club in 2009 and 2010. A student or two learned to pick locks, a skill of necessity when late-night tinkering required something that was not at hand but could be cannibalized from a piece of hardware on the other side of a door.

Raphael Jedidiah Sofaer was the first of the family not born in New York City. He had been aiming toward the city throughout high school. The family had moved to Palo Alto when his father was appointed as a fellow in the Herbert Hoover Institute at Stanford. The family home in Palo Alto, California, often hosted prominent government figures, and the Sofaer sons were always ready to debate. Rafi, baby-faced, was the most level-headed of the Diaspora group, eager to discourage overblown expectations of what they were going to accomplish, but also to spell out its necessity.

Before NYU, Rafi went to the Woodside Priory, a small independent boarding school founded by Benedictine friars near his family's home. He spent summers at Hebrew camp. Hacking was not the center of his life, but a way to scratch his intellectual and political itches. Slight in build, short in height, and an acute listener, he would sit in a room filled with raging argument, a half-smile laced to his face. No one would mistake him for the stereotypical math whiz, corralled and isolated by his intellect and drawing satisfaction only from numbers; though not a practitioner of small talk, he was open and friendly.

Going to New York had been Rafi's dream, but it was a long distance from the cozy atmosphere of the friary. And NYU itself is an unruly village that sprawls across square miles of lower Manhattan in patches from the financial district to Union Square. Many dormitories are a subway or bus ride from the core of classrooms near Washington Square Park. Students lug books and laptops for hours. Everyone improvised places to take breaks, preferably somewhere they could just put their stuff down.

The ACM room on the third floor of the Courant Institute was only a few hundred square feet, with five or six computers, but it offered a psychic niche, a hangout between classes. Rafi had been looking for a kind of nest, or safe harbor, and he found one in the ACM room.

At twenty-two, a few months from graduating, Max seemed to have found his footing in the tech world. He had come to NYU with wide-eyed dreams of being a musicologist, and went to work as a radio DJ on the campus Internet station, then took a part-time job with an independent record company. By the end of two years, the culture of nonchalance in the company, the all-purpose dismissive "whatever" that he heard when he pushed to get music out the door, had battered his spirit. "Nobody likes the enthusiastic music kid," he said later. "I want to be in a situation to succeed, and have that be a good thing—and not to be shunned because of that."

Born in 1987, Max had been reared, like his classmates, in the digital age. He was fluent from an early age in computer programming. His father had spent much of his life in retail food and beverage stores, having worked for the Campbell Soup Company and Gallo wines. As a boy, Max had tagged along on his dad's visits to grocery stores, and took in the thinking behind their strict layouts: fresh foods like vegetables, meat, and dairy along the perimeters, and categories like soups, soda, and soaps in the center. They were completely engineered spaces, he thought, much like websites.

His capacious curiosity took him to a course in the anthropology of hackers, and there he became certain that the technology world would welcome his wholeheartedness. "In Silicon Valley, the enthusiastic kid who has the crazy idea gets very lucky," Max would say. He was also dogged: good at completing a task. The previous September, he enrolled in a class on heuristics, strategies for solving problems. After the first meeting, he was approached by a slightly disheveled, smiling student. His name was Ilya Zhitomirskiy, and he had noticed Max wearing a button with the logo of the Electronic Frontier Foundation, a legal nonprofit that fought to keep civil liberties pinned into place as technology created new platforms for speech and commerce that had not been anticipated by centuries of legal precedent and law. Ilya was impressed.

"I really like your pin," he said.

"Do you even know what it is?" Max asked.

Ilya snorted. "Of course I know what it is," he said.

Officially, Ilya was not taking the heuristics class but dropped in because he had heard, as Max had, that the subject was hard and the professor interesting. That curiosity and fearlessness proved a strong adhesive. They had no doubt about the centrality of programming to life in the twenty-first century.

Ilya dropped in on another one of Max's classes, taught by his adviser, Gabriella Coleman. She had asked the class a question. What happens when people are being watched? No one had the answer, but the question she posed and then answered herself, stayed with Ilya.

When people are being watched, she had said, they perform. Ilya loved that insight. He was an expert at lighting the stage, literally.

Two years younger than Max, a puppy in spirit, Ilya was already well into PhD-level math courses. His grandfather was a mathematician. So was his father. When he was twelve, the family moved from Orekhovo-Zuevo, a town outside Moscow, to the United States, and began their American life in New Orleans. Ilya was decidedly off-kilter, nervous about fitting in, but saw no reason that he shouldn't wear pants in the brightest neon colors that he could find, or giant plastic orange sunglasses.

At his first school in the new country, he was astounded by the familiarities of the teachers, with their cheery greetings of "Hi!" They had none of the sonorous formalities of the teachers in Russia who had taught him bits of the queen's English. One day the principal appeared at his classroom door and summoned Ilya to come with him to the office.

What, Ilya fretted, had he done?

At the office door, the principal turned to him. "I can't get into my computer," he said. "I lost the password. They tell me that you might be able to fix it."

Five minutes later, the principal was back on his computer. Ilya was rewarded, to his amazement, with a one-hundred-dollar gift certificate. Still, he managed to run into a bewildering string of trouble with some Americans. The family moved to Boston from New Orleans, and he was not yet sure-footed in English. Another boy from Russia often translated for him. After weeks of this, Ilya realized that the other boy had been entertaining himself by deliberately warping the translation and watching Ilya squirm.

By junior year in high school, Ilya had full command of English, and the Zhitomirskiys had settled in Lower Merion, Pennsylvania, just outside Philadelphia. His Russian accent had been sanded down by four years in the United States. He decided to leverage his perennial newcomer act into a social project. Every day, he would meet a new person. Just walk up to someone he didn't know and, in the easy American way, say, "Hi, I'm Ilya." The trick was making it seem casual. In truth, before he approached strangers, the anxiety sweated into his palms, and he would have to dry them on his sleeves. He was one of the few kids in school who did not have a Facebook account. Instead, he built his own social network, one handshake at a time. Before the year was out, he was working with the stage crew on the school drama club productions, happy to dry his palms and pretend that climbing the high rails to adjust the lighting was no big deal. In the summer before his senior year, he knocked on the stage door of every theater in Philadelphia, looking for work. Few people had time for him but he kept going.

At the Academy of Music, which was showing *The Lion King,* the manager said that he had no jobs, that Disney did all the hiring, and that Ilya wouldn't be hired in any event because he wasn't a union member.

Another rejection.

But then: "Do you want to come in and see how things work?" the manager asked.

Soon, Ilya was striding along catwalks in the upper reaches of the hall. Eventually, he landed tech jobs at a few small theaters. For college, he wanted a top-ranked math program, and was accepted to one at the University of Maryland, though his seat would not be available until the January after his graduation. That meant he had a term to kill, so he registered at Tulane in New Orleans, where he joined the juggling team and learned to unicycle. Following that semester, he transferred to the University of Maryland in College Park, and after math classes took up competitive swing dancing. He also power-kited, propelling himself over the crests of hills and gliding as far as he could. For meals, he practiced Dumpster diving—salvaging edible sandwiches and even sushi from the bins outside a coffee shop. The Maryland program was strong, but he was drawn to a place he had never lived: the heart of a big city. The Courant Institute of Mathematical Sciences at New York University, housed in Greenwich Village, piqued his academic and personal interests. So for the third time in three years, he began at a new college.

It was as if he had landed in a black-and-white photograph of the mythic, shimmering metropolis. His first dorm was across from the Brooklyn Bridge; his roommate was a jazz musician who came home from gigs at three in the morning and played a keyboard while Ilya worked on math problems. In the lounge of the math department, he would laughingly harangue students who spent time on Facebook; why, he wondered, were they wasting time with such fake relationships, where people just spread themselves out? Where was the joy of discovery? One day, as a prank, a classmate set up a Facebook account in Ilya's name and friended everyone in their circle, including other students and the graduate teaching assistants. When he discovered the trick, it was the only time most people saw him truly angry. He confronted another student, Stephanie Lewkiewicz, with whom he had developed a romantic relationship. She conceded knowing that the scheme was afoot, but swore that she had not instigated it or taken part. Another student had.

"That guy friended all my TAs," Ilya sputtered. He was shutting down the account. "What are they going to think? Everyone is going to be offended thinking I just defriended them."

"But honestly," Stephanie said, "I didn't actually do it."

"I believe you that you didn't do it. You probably just thought it was hysterical," Ilya said.

"I did," she said.

He hunted for the Facebook account cancellation procedure, but had no faith that it would actually work. Somewhere, he was certain, his personal data was going to be stashed on a server.

"They're not going to let me really delete it," he told Stephanie. "It's going to be stored on the Internet. It's never going to go away."

That single episode aside, he was a bright spirit among the muted tones of the math program. He often wore a shirt designed as an American flag and neon blue pants; he sat near the front, hand up, more with questions than with answers. His wandering curiosity brought him to Max's heuristics class, and then to the computer club room, which had a much livelier hangout scene than the math lounge.

As it happened, Max and another senior, Dan Grippi, were officers of the ACM, and they had an intuitively subversive approach to computing. Or perhaps they saw it as unmapped land, mostly ungoverned. They turned the club office into a hangout through a satisfying series of small hacks. Until their administration at the start of the senior year, it was hard for anyone to just drop in because only the club officers had keys, but Max and Dan fixed that. Not by going to a locksmith for more keys. One night, they put a little radio frequency identification chip on the door. The chip, known as an RFID, is a simple gadget that sends signals a few yards; it's what lets an identification card swipe open a turnstile, or lets a driver pay a highway toll without cash. Max and Dan set up their RFID to indicate when the door was opened or closed. The signal was strong enough to reach the computers in the room. Then the computer would send out a tweet under a Twitter account registered as @acmroom.

So the door had its own Twitter account. It tweeted simple messages.

The ACM Room is open.

Or:

The ACM Room is locked.

To accomplish that, however, Dan and Max had to make changes on one of the computers in the room, which ordinarily could be done only by people with the master password privileges, like the university's systems administrators. The proper way to do it would have been to file a work request ticket through channels. Or they could just hack into it.

With everyone gone for the night, Dan read out the serial numbers from the back of the computer, and Max pulled up the manual from the Dell website. It explained how to reset the administrative password. You open the computer casing and unplug a jumper that supplies power to the motherboard, then waited a few minutes. Then you can put in any master password. Easy.

The door's first tweet had said:

if a door opens and no one is there to see it, is it really open?

Message number 2:

the ACM room door is now closed!

On his own Twitter account, Max had announced:

sweet, now all of you people can know when I am in the acmroom because i made this door tweet.

Even with glitches, it was a cool stunt that won new fans for the room and the club.

For a partner in a minor black bag operation, Max could hardly have done better than Dan Grippi. Patient, methodical, inscrutably quiet, Dan had been elbows deep in computers since he was five or six years old.

During his junior year, he had applied for work at StyleCaster, a website created by two brothers who had graduated from NYU. At the time, StyleCaster offered suggestions on what people should wear that day, based on the weather. During the job interview, Dan noticed that they had a big photo studio. Just for fun, he wrote a short programming script on the spot that automated part of the photographer's work.

If he were hired, the interviewer asked, how much would he want? Dan wasn't sure.

"How about eight dollars an hour," the interviewer suggested.

"No way," Dan said.

"Well, what do you want?"

Dan did not have a figure in mind, but wondered if he could get away with asking for fifteen dollars. Maybe they would go for twelve dollars. He often thought before he spoke. Before he could get the number out of his mouth, the interviewer answered his own question.

"How about twenty dollars an hour?" the interviewer said.

"Sure," Dan said.

Not for the first time, his taciturnity had paid off.

After he and Max took over the computer club, kids were in and out of the place at all hours.

One day, signs appeared around the computer science building:

**THURSDAY NIGHTS
COME AND GEEK OUT!
BUILD A MAKERBOT
EAT PIZZA
BRING YOUR SLEEPING BAGS.**

The MakerBot is a three-dimensional printer, a contraption of transformational possibilities. It is a machine that makes things. The 3-D printer works like an inkjet printer except that it squirts molten plastic, not ink. As layers of plastic are added, they can rise into almost any shape—bolts, tools, toys, knobs, bullets—based on digital designs sent to the printer from a computer. In short, it was a factory for the desktop. Already, industrial versions of 3-D printers were making exotic furniture for hotels and restaurants, lamps, doorknobs, jewelry, handbags, perfume bottles, clothing, and architectural models; the television personality Jay Leno had one that created custom part models for his vintage automobile collection, keeping a 1907 steam-driven car on the road. The ready-made commercial 3-D printers cost from $25,000 to more than $1 million. All things considered, the notion that a bunch of college kids would be messing with a 3-D printer would have been beyond conception a few years before. But earlier in 2004, members of a hacker collective in Brooklyn had started selling MakerBot kits, the electronic guts and hardware of an assemble-it-yourself 3-D printer. The cost:

seven hundred dollars. After it was put together, users could download ready-made designs from the Thingiverse, a repository that anyone was allowed to use or contribute to. It was the perfect teething biscuit for the crowd in the ACM room, decided Evan Korth, the faculty adviser to the group.

Korth had spent a good part of his ten years as a faculty member hunting for new trails into the world for NYU's brilliant math and computer science students. The easy, obvious route led to Wall Street, which had an insatiable appetite for programming whizzes. But Korth wanted the students to see that not all paths led to the investment banks, where so many of them would end up twirling algorithms in the financial services rodeo. With Hilary Mason, a data scientist from Bit.ly and Chris Wiggins, a physics and math professor at Columbia University, Korth started a nonprofit called HackNY. Its unofficial slogan was "Save the kids from the Street"—meaning Wall Street. They staged hackathons, twenty-four-hour competitions for teams to hack together a software program that did something fun.

Korth did not begrudge his students the respectability of sweet-paying internships, but he wanted them to also see the possibilities in hacker spaces, the lofts and storefronts occupied by cooperatives of tinkerers. His students were the first generation for whom it was simply intuitive that power was exercised in 1s and 0s. They appreciated technology, but neither feared nor revered it. The most adventurous of them opened the backs of phones and appliances, heedless of the perils of voided warranties, to see what new, unplanned tricks the gadgets could be made to do.

Korth brought the ACM group to Brooklyn to visit NYC Resistor, a hacker space up four flights of steep wooden steps on the top floor of a nineteenth-century building that had once been the home of the Long Island Brewing Company.

"NYC Resistor is a force of chaos in the world," said Bre Pettis, a member of the collective that started the hacker space. "Imagine kittens with jet packs—flying around in a room."

It was there that the MakerBot had been conceived by Pettis and a group of hackers who thought that 3-D printing was exciting and that nothing about the technology warranted a price tag of tens of thousands of dollars. (In 2013, MakerBot, with backing from the venture capital arm of Jeff Bezos, was sold to an industrial maker of 3-D printers for $403 million.) They had started their business creating build-it-yourself kits and selling them for less than one thousand dollars apiece; with a few extra dollars in the budget, Korth bought the forty-sixth one produced. He invited the two officers, Max and Dan, to figure out something to do with it.

The Association for Computer Machinery began in 1947, after modern computing began to grow from concepts developed by Alan Turing and John von Neumann before and during World War II.

The work of Turing and von Neumann "broke the distinction between numbers that *mean* things and numbers that do things," the historian George Dyson wrote. "Our universe would never be the same."

Almost seven decades later, college students could try to build their own desktop factory. Max posted the signs inviting people to geek out and build the MakerBot.

The early versions of the MakerBot kit required lots of assembling of hardware parts. Ilya showed up the first night with his own soldering kit, which was supplemented with an electric hot plate for baking parts. To make sure that no one would mistake this cooking appliance for something to actually cook on, one of the students wrote a felt-tipped warning on the lid: "No Food! Lead Only!" A cheery-looking skull and crossbones emphasized the point.

The building sessions went on for weeks, late into the night. Korth, who lived not far from the school, usually stopped by with snacks to fortify them. One time he brought a date along. Afterward, the woman asked: "Who was the kid in the neon blue pants? He was hitting on me like crazy." That was Ilya, Korth realized, laughing at the audacity.

As the semester was coming to a close, they had not quite finished assembling the bot. Max was determined to see the project through, and the group agreed to gather for a final push on the last Saturday of the term. They hooked up a camera to document progress on the contraption, and streamed video live to the World Wide Web. The revolution was being televised. Korth, at a party, spent most of the time in the host's bedroom watching the video feed on his computer.

After going at it hard for an hour or two, Ilya's stomach was growling. There had been no pizza that night. His voice rose from the box.

"Can we get some food?" he asked.

Max broke the news. "We have some crackers," he said. "Let's finish the food we have. We're running out of money."

Korth left the party and headed for an all-night deli to pick up ice cream for them. Dan noticed that the official MakerBot site was tweeting updates on their progress to Twitter, and also sharing a link to the live stream. "Over two thousand people are following MakerBot, so they're going to see what we're doing!" he shouted.

From his perch inside the box, Ilya gave a small sigh of satisfaction as he got a tricky piece connected. "There is something to be said for having the right tools," he declared.

"And that thing," Rafi said, "is 'awesome.'"

Ilya dug into the ice cream. Rafi pushed a button on his laptop, which loaded the design instructions and sent them to the printer. The first layer of plastic was squirted precisely into place. It would take an hour or so, but eventually their first creation would take shape: a plastic shot glass, the traditional product to mark the inaugural of a new MakerBot.

The shot glass came off the printer near four A.M. A credible result, if a bit lopsided. They poured Maker's Mark. Dan and Max, Ilya and Rafi, were elated. College students are known to like a drink, but not for making their own shot glasses, much less building a machine to custom-manufacture them.

It was worth toasting, and they did. By the start of the second term, they were looking for something new to do together. The long nights with the MakerBot had baked the four guys together. So when Korth told them that the Internet Society was sponsoring a talk by Eben Moglen on a Friday night in February, they had been ready to hear him say that the world needed fixing and that they could start with social networking. There was indeed something to be said for having the right tools.

# CHAPTER THREE

**D**an was in agony after Moglen's talk. He had not gone to it, as he was backed up with homework, due to graduate in May, and feeling some parental pressure to find a job. So he stayed in his apartment in the East Village. But the other dudes were there, and he wanted to at least be conversant. So he had surfed to the webcast of the speech, and he set it as his aural wallpaper, half listening, then three-quarters, and then stopped everything else.

It was a few nights after the talk that Dan started peeling back the settings on his Facebook account, curious to see if he could find a way to regulate his privacy.

He could not. When Dan saw the guys in the ACM room, he told them that he had quit Facebook, and was going through withdrawal. They really had to build a better social network. In bull sessions there and elsewhere, Max and Ilya, Dan and Rafi, and a few other friends talked about Moglen's assertion that Facebook was just a collection of PHP doodads, with spying thrown in for free.

The running updates from friends? "There's nothing unique about activity streams," Rafi said.

Pictures?

"We've been putting pictures on the web since 2000," Ilya said.

If nothing about Facebook was technically exotic, the elements were all pulled together so crisply that hundreds of millions of people had becoming dead cheap. Virtually all the essential software was now open source and free. Computers got cheaper and more powerful every few months. In fact, for the mature technology companies, often lumbering masses no longer recognizable from their sprightly beginnings, it was less expensive and simpler just to buy promising start-ups than to create new products themselves or spend a fortune recruiting someone.

Max had read essays by an entrepreneur named Paul Graham, who created a program that allowed users to build their own Internet stores. Yahoo! bought it for stock worth nearly $50 million.

Graham argued that the old paradigms of venture capital had shifted. Entrepreneurs did not need to go to business school as long as they knew computer science; learning to program was the hard part. Max was persuaded. By sophomore year, he decided to double major, adding computer science to his communications studies. The world Graham described sounded infinitely more attractive than the ossified music company he'd worked at. In fact, Max was supposed to be writing part of his senior thesis about Graham.

If young people had good ideas, Graham had written, they could take them directly to the public, without risking slow suffocation inside giant companies.

On the Internet, nobody knows you're a dog. And more to the point, "nobody knows you're 22," Graham had written. "All users care about is whether your site or software gives them what they want. They don't care if the person behind it is a high school kid."

Moreover, a common condition of twenty-two-year-olds—being broke—no longer mattered, given that the cost of starting a tech company had almost vanished. "It's so low that it has disappeared into the noise," Graham wrote. "The main cost of starting a Web-based startup is food and rent. Which means it doesn't cost much more to start a company than to be a total slacker. You can probably start a startup on ten thousand dollars of seed funding, if you're prepared to live on ramen."

If you could start an actual company for no more than it cost to hang around in your mom's basement playing video games, how much would they really need for a summer project? They could survive on boiled water and noodles in a cup.

Once again, they were staying late in the ACM room, writing lines of code. One night, Ilya arrived with a razor and a comb; he would work late, then hit the couch for a few hours before rising for a rudimentary wash before heading off to class. Rafi was horrified. As the work rolled on, he had a choice between dropping out of classes or dropping his social life. He stayed with the project.

Walking along Fourteenth Street, Max glanced at the papers on a newsstand. A word from a headline stuck in his head. He carried it back to the ACM room, and proposed it as the name of their project.

"Diaspora," Max announced.

"What does it mean?" Dan asked.

"I never heard of it," Ilya said.

Rafi knew it well. Beginning with the exile of the Jews from Babylon, the word had been used to describe a community that had been widely dispersed. To Max, it reflected the scale of their ambitions. It wasn't just about Facebook. It was like restoring the world.

Dan was indifferent about the proposed name. It might not have been a word everyone would recognize, he thought, but at least they could call the project something other than "That thing that we're trying to build."

Ilya looked up the word, saw that it originated from the Greek for scattering of seeds.

"It's perfect," he said.

They coded through the night: headphones on, each one absorbed in building the ladder of logic needed for a function to work.

By the end of the month, they faced the first adult test of their scheme. They had to get money for it—not for that very moment, while they were still in school, but for three months down the road. As it happened, Paul Graham, the subject of Max's unwritten thesis, ran a program for startups called Y Combinator. Graham had created a seed fund that would offer up to twenty thousand dollars and three months of advice. Twice a year, Y Combinator accepted applications. Not only were business plans not required, Graham pointedly said that they would not be read. Once they were accepted into a class, the new ventures were required to move, at least temporarily, to Silicon Valley. For three months, they would have regular skull sessions with the Y Combinator partners. They also had to meet once a week with big names in the tech world at off-the-record dinners. (Its name was wordplay familiar to those who had wandered deep in the arithmetical forest: the Y combinator is a mathematical function that generates other functions, and Y Combinator was a company that generated other companies.)

The next cycle of the Y Combinator camp would run from June through August, and the application was due at the beginning of March. The Diaspora guys gathered on the final Sunday in February to work on it.

Name. Company name. Phone number. Link to a one-minute video introducing the founders.

They did have a company name, although they were hardly a company. They were four nerds. Or, when their self-esteem was cresting, four dudes.

Please tell us about the time you most successfully hacked some (non-computer) system to your advantage. Please tell us in one or two sentences about the most impressive thing other than this start-up that each founder has built or achieved. How long have the founders known one another and how did you meet? Have any of the founders not met in person?

The application form was a blizzard of provocative questions, twenty-seven blank spaces to fill in, most of them requiring responses to three or four deep queries. As a whole, the inquiries were perfectly reasonable, considering that they were designed to figure out if it was a good idea to hand over twenty thousand dollars to complete strangers with little to no history of running a business.

How far along are you? Do you have a beta yet? If not, when will you? Are you launched? If so, how many users do you have? Do you have revenue? If so, how much? If you're launched, what is your monthly growth rate (in users or revenue or both)? If you're not incorporated yet, please list the percent of the company you plan to give each founder, and anyone else you plan to give stock to. (This question is as much for you as for us.)

"Man, all these questions," Max said. "We're four smart kids."

Please tell us something surprising or amusing that one of you has discovered. (The answer need not be related to your project.)

There were easily another fifty questions along those lines. Within a few minutes of looking over the application, they bagged it. June was still three months off. Something might come up. In the meantime, they would continue writing code. Unlike the application, it was too interesting a problem to let go of.

In February 1997, a piece of dismal news reached the online message board of the Freaks, the fans of a British and Irish progressive rock band known as Marillion. Their record label had gone broke. The keyboard player, Mark Ryan, wrote that the band would not be touring North America with their new release.

In Bedford, Massachusetts, this seemed outlandish to Jeff Pelletier, a fan of Marillion. They might not have been the Rolling Stones, but plenty of people were devoted to the group; the bulletin board hummed with news and conversation about it. So Pelletier put the idea to the online Freaks: why don't we raise the money so they can tour? The band members were dubious, and said it would take thirty thousand dollars to cover their touring costs. Even so, they assigned a friend to oversee whatever money did arrive so that the well-meant efforts of Pelletier did not turn into a fiasco.

In short order, small contributions began to roll in, from all around the world. "People were giving money just to see this happen, because it was such an amazing thing," Pelletier later told the *Chicago Tribune*. All told, the Freaks raised about fifty thousand dollars. Marillion toured triumphantly, rewarding their backers with autographed DVDs. The market, or the street, got to vote.

Five years later, Perry Chan, who had worked as a day trader and a nursery school teacher, tried to put on a concert during the New Orleans Jazz Festival, but could not risk twenty thousand dollars of his own money. There had to be a way to round up people online who supported good music. He moved back home to New York and found a job in a Brooklyn diner. Wouldn't it be good, he would occasionally muse out loud, if there were a way to use the web for lots of people to provide small-scale backing for creative projects? One of his regulars, Yancey Strickler, a music critic, thought it was a fine idea.

With Charles Adler, a graphic designer, Chan and Strickler created Kickstarter, a website where people could pitch projects and perhaps collect enough small change to launch them.

The setup was simple and ingenious: the artists or inventors could post a description of what they were working on, how long it was going to take them, and how much money they needed to raise. The pledges could be made to Kickstarter through a payment system operated by Amazon; the website kept 5 percent. A strict rule gave the fund-raising drives a thrilling, or harrowing, edge: no money would be distributed unless the goal had been met in full by the deadline. It was all or nothing. If people said they needed five thousand dollars to do their project, and they got only thirty-five hundred dollars, well, that wasn't enough. In that case, the donors would get their money back. The requirement was to drive those seeking support to set realistic goals, and to hustle for contributions.

Kickstarter permitted the drives to go on for ninety days, but a performer or inventor could set an earlier cutoff, if, for example, a show was a few weeks off.

Artists typically used Kickstarter to raise a few thousand dollars or so for videos and comic books and music. So did people like the ecoentrepreneur with plans to rejigger old gum-ball machines that would dispense wildflower-seed bombs; a graphic designer who proposed to fly cameras on kites and balloons along the coastline to create intimate, bird's-eye maps of the big 2010 oil spill in the Gulf of Mexico; and the young cook in Kansas City who hoped to fix up a trailer so she could expand her sales of 100-percent-natural snow cones. In short, Kickstarter was a way for the far-flung to find and support the far-fetched, the charming, and the essential. It was distant from the kinds of gift giving that result in concert halls and libraries being rebuilt and renamed—the megaphilanthropy that could be seen as an expression of love for humanity, or the graffiti tags of wealth.

One early project backed through Kickstarter was a hackathon run by HackNY, a daylong binge of coding, pizza, and Red Bull, a scheme hatched by Korth and his friends to connect college kids with start-up tech companies.

Teams of students from all over metropolitan New York would have twenty-four hours to hack together a cool project, then demonstrate it. To feed and water the hackers, Korth and company went to Kickstarter for $2,000. They got $5,005.

Convened at the beginning of April, the hackathon was a roaring success, with delegates from schools all around the city.

The Diaspora guys, of course, were there; the event was being held in the building where they had spent most waking moments since abandoning their application for the Y Combinator camp. They had written swaths of code to create the frame for Diaspora, and were starting to build in essential social network functions, but they were miles from being done. So they came to the hackathon with no notion of having a clever app to display after twenty-four hours. Rather than chop off some little piece and put it into the hackathon competition, they seized a chalkboard and worked as if they were in the ACM room, continuing to tunnel toward Diaspora, amiably answering questions from people strolling around.

Fred Benenson, who had spoken with Max after the Moglen talk, spotted them and came over to chat. He had recently started working for Kickstarter and was impressed by the progress they had made in the two months since the speech.

"This could work on Kickstarter," Benenson said.

That thought had crossed their minds. In fact, Korth had mentioned it to them as a possibility.

"E-mail me," Benenson told Max.

They soldiered on through the twenty-four-hour hackathon, taking a break at two A.M. when Korth and Hilary Mason showed up with ice cream and toothbrush kits. They ate bagels at dawn, and hung around until noon to watch other teams collect prizes. Then Max charged off to write to Benenson, putting the skeleton of the idea into an e-mail. A social network built around privacy. You owned your data. Decentralized.

Fred loved it.

The calendar had to be taken into consideration. Although Kickstarter allowed up to ninety days to collect contributions, this was a summer project, so they had less than two months to raise money.

Before they got to that problem, though, they had to find out if Kickstarter would even let them post. Certainly, it helped to have someone on the inside, like Fred, pulling for them, but real estate on the website was valuable. Benenson gave the matter some thought. He could simply extend an invitation to Diaspora—Max, after all, was his friend—which would practically guarantee the guys a spot, but it seemed that going through the Kickstarter review process would sharpen their presentation. The guidelines broadly required that the pitches be for projects that had some creative purpose, like making a film or a new game, and a clear end point. It was not a place to raise money for overextended credit cards.

Fred coached Max and Dan. The project had to be clearly, concisely explained for the website. They should also make a video introducing themselves and the project; it would give people a chance to connect with them as humans, not just concepts. It was one more thing to do for the end of the term. Max had a senior thesis to write for his favorite teacher and adviser, Biella Coleman, the anthropologist who introduced him to the study of hacker culture. Dan was thinking about jobs. And their nights in the ACM room, grinding away on Diaspora, were getting longer and later.

On April 15, Max and Dan simultaneously started getting messages from other students that they had better get over to the ACM room, fast. Max was blocks away in his apartment, but Dan was nearby, and he arrived in a few minutes.

All the computers were gone. One of the systems administrators had come down and seized them because the university network had not registered any activity on the machines in the room for months. They had virtually disappeared from the network. The systems administration wanted to know how that could be.

"I was one of the people who broke into them," Dan said, explaining that they had hacked into them so the door could have a Twitter account.

"So," the administrator said. "You thought this was a good idea."

Dan considered this formulation. They had wanted to make them better. Filing a service ticket and waiting for however long it would take to get the system updated did not make a lot of sense.

"We weren't doing anything bad," Dan said. "We were just exploring the machines."

"So you thought this was a good idea," the administrator repeated.

It seemed that the man wanted Dan to say those words.

"Really, we didn't do anything malicious," he said. "When are we going to get the computers back?"

Negotiations ensued. Korth, the adviser to the ACM club, was brought in as an intermediary. The settlement was a letter of apology, which only the seniors signed, under a tacit understanding that they would take the blame, such as it was, for breaking into the machines and making them work better. The computer club really did need computers in the club room.

A final tweet was issued by the ACM Room Door account:

@acmroom

Just lost the game.

Diaspora owned them. Besides coding all night and hustling to get the Kickstarter appeal in shape, it turned out that they had to come up with premiums for people who contributed money. What the heck was this, a public radio fund-raiser? Stickers they could manage pretty easily. T-shirts for people who gave a bit more. Benenson had a suggestion for the midlevel contributors: as a kind of joke, they could give out CDs with whatever code they came up with by the end of the summer, signed by them as if they were rock stars. For megadonors of two thousand dollars, they would give computers loaded with Diaspora and a promise of easy support. These were outsize promises, but the only ones they expected to donate anything like two thousand dollars were their parents, who were unlikely to fuss if they didn't get a new computer from their sons as a fund-raising premium.

Korth suggested that they give a presentation on Diaspora to the Internet Society, since the project was inspired by the talk Moglen gave at one of the group's meetings. It was a freewheeling session.

Later that night, they had to make the video for Kickstarter. The four of them sat in a classroom, writing discussion points on a blackboard. Who are we? What is it? Why does it matter? Their dress was motley, but they came by it honestly, not in the practiced manner of hipsters: a few

T-shirts, a bowling shirt, some stubbled chin, a few coffee cup accessories. They sat behind a teacher's desk. No cameraperson: Max perched a pocket-sized video camera on top of an overhead projector.

After introducing themselves, the stilted recitations began.

Then Dan jumped in.

"Hold on," he said. "What is this project? Because I have absolutely no idea."

Cue to Rafi: "In real life, we talk to each other. We don't need to hand our messages to hubs and have them hand it to our friends. Our virtual lives should work the same way."

He spoke in robotic bursts: "We don't know what's going to happen to our data. It's going to exist into the foreseeable future. So—[he seemed a little flustered]—we need to take control of it."

Max broke in: "Because once you give it away once, it's no longer yours. You cannot stake claim to it."

"That's why my mom won't put anything on the Internet unless she's ready to see it in a newspaper," Rafi said.

"But we know that's bogus, because sharing is a human value," Max said. "Sharing is what makes the Internet really awesome. I think it's what gets us all excited whether it's a silly little LOL cat or your newborn nephew—"

"My newborn kid," Dan said, drawing a giggle.

Ilya, whose only contribution up to that point had been warm smiles, said: "No longer will you be at the whim of these large corporate networks, who want to tell you that sharing and privacy are mutually exclusive. Because it will be your node." No one uttered the word "Facebook" in the video. Ilya's comment was the closest they came.

At one point, they seemed to be at a loss, but Dan spun around and moved one of the sliding blackboards to find the next line.

"Why us?" he said, turning back to face the camera.

"So," he said. "Why us?"

Rafi jumped in: "Us—because we're ready to do it. We're ready to work full-time."

"Twelve hours a day," Ilya said.

"I think it's a little more than twelve," Dan said.

"We're ready to give up three months of our lives," Max said. "Dan and I are graduating from school. Ilya and Rafi turned down some sweet internships. We're really passionate about this idea. We really want to see it, so we can use it."

Max would later estimate that the video was the single most awkward moment of his life. But it worked. They were so obviously struggling that one blogger wrote, "After watching the video of students explaining their idea, saying 'no' would be like turning away a Girl Scout cookie seller empty-handed. We just don't have it in us."

They'd had a budget of twenty thousand dollars in their heads, the product not of any sharp-penciled calculations of what it would cost them to live but simply because it was the maximum amount offered by the Y Combinator program that they had ended up not applying for. The most successful Kickstarter project to date had raised more than eighty thousand dollars, but nearly all the others had brought in less than five thousand.

Korth argued that the twenty-thousand-dollar target was a mistake. Before going into teaching full-time, he had a life in the business world, working as a sports agent and at a successful web company. They needed to crunch the numbers tighter.

"Remember," he said. "It's all or nothing. Set the goal too high and you could end up with nothing."

If they stuck with the plan of camping out in the Salzberg house, their housing expenses would be minimal. They'd still need round-trip tickets to Lake Tahoe, some kind of transportation while they were there, gas, and food for three months. Maybe some computer server space.

Could they do it for ten thousand dollars?

More important, was it realistic to think they could raise ten thousand dollars in contributions? Their families would probably pitch in, maybe some aunts and uncles. Their friends were, like them, college students with no cash to spare for philanthropy. It was a lot of money.

As they got closer to pushing the button, the risks of winding up with nothing got to Dan. He was blowing off job interviews. He called Max.

"Maybe we should go for eight thousand dollars," Dan said.

Max tried to calm him down. They could get their families to top off if they were close.

The pitch went live on Friday, April 23. First across the line, no surprise, was Evan Korth, who pledged one hundred dollars.

Two other friends pitched in, too. But that was it. One day, three contributions.

This time, Max called Dan.

"We should have gone for eight thousand dollars," Max fretted.

# CHAPTER FOUR

**B**y Sunday night, they had gotten donations from twenty people, including Korth, Fred Benenson, and Max's sister Lili. Not bad, but not quite a land rush. A few donors mentioned on the comments page that they had been thinking about precisely this kind of project. The guys were checking in all weekend, to see how it was advancing, and they were delighted to trade shoptalk with the fraternity of geeks who had also thought about the subject.

Late Sunday evening, a peculiar e-mail arrived, not through the Kickstarter website but at the account they had set up for the Diaspora blog.

"I'm just writing to thank you for creating this project.

"My name is Yosem Companys, and I am a PhD candidate in engineering at Stanford University. I had thought about your very project around two years ago and have spent just as long evangelizing about it to anyone who would listen in the hope that someone would take the initiative and do it. I wish I could get involved, but I just don't have the bandwidth right now with all my other doctoral obligations."

He went on to mention that he was involved with a technology ventures program at Stanford, and would be glad to connect them with folks who might be willing to help with advice, money, etc. "So if you need any help, please don't be afraid to ask," he wrote.

Max quickly sent back a note of thanks.

"Right now we are in fund-raiser mode for our Kickstarter, so anyone you may know that would be willing to show their support or enthusiasm to help us get the project started would always be appreciated," he wrote.

After the summer, Max wrote, there would probably be more opportunities for the venture community to help them continue the project. He signed off by saying that Diaspora could be followed on Twitter and through their blog. The whole exchange seemed preposterous. They hadn't put a line of code in public, and this guy was hinting that he could fix them up with venture capital? Max put the matter out of his mind until the next morning. Then more e-mail arrived from Yosem Companys. This time it was a copy of messages he had sent to senior partners at two venture capital firms, discussing a wonderful start-up opportunity called Diaspora, and, by the way, his brain surgery had gone well. Max shook his head. Brain surgery. It was starting to make sense.

An hour later, they got a message from one of Yosem's correspondents, a man named Randy Komisar. "I love the Diaspora idea," he wrote. "Believe it or not, I have been working up a related concept. This is important. We need control of our information. We may share it, trade it, publish it, but it should be 'we' who do it.

"Max, Dan, Raphael, and Ilya, I would love to meet. Maybe even tomorrow."

A few seconds on the Internet introduced Randy Komisar: a former CEO of LucasArts Entertainment, the video game division of the empire of George Lucas, the director of *Star Wars*. (Lucas had turned down his director's fee on the first film in favor of licensing rights to the toys and games, deemed to be worthless by Fox Studios; the sales would allow him to self-finance the sequels.) Komisar had also been a senior counsel at Apple, and had written a bestselling book on starting businesses, *The Monk and the Riddle*.

But it wasn't just Komisar. It was what and who he stood for. He was a partner in Kleiner Perkins Caufield Byers, a venture capital firm in Silicon Valley that had been an early backer of Amazon, Google, Lotus, Sun Microsystems. They were Goliaths.

Max had to read the e-mail several times. So did Rafi, Dan, and Ilya, all of them trying to get their heads around the news. It could all be a hoax, someone spoofing them. Or had this guy Yosem actually gotten one of the biggest venture capital firms in Silicon Valley to contact them about their project?

Not only that, this Yosem had sent e-mails to twenty or thirty other famous names in technology, government, and finance.

"Forgive me," Max wrote to Yosem, "because I am a little taken aback about how awesome this list is."

They would get back to Randy by the end of the day.

Another e-mail arrived from a lawyer, Mark Padilla, with Wilson Sonsini Goodrich & Rosati, a law firm in Palo Alto that specialized in start-ups. Rafi knew them well: his brother Joseph had been involved in a start-up that had used the firm. He wrote back to set up a conference call.

Dan phoned his mother to report on all the people Yosem Companys had contacted.

"Guess who else this guy e-mailed it to," Dan said. "Obama's technology adviser!"

"That's nice," his mom said. Carolyn Grippi was the chief operating officer of the Daytime Emmy Awards program, and both she and her husband, Casey, an accountant who was about to retire from Citibank, were skeptical of this plan to build a social networking site rather than get a job. They had just paid for four years at one of the most expensive private colleges in the country, and their son was talking about living on noodles?

Then again, she thought, if not now, when?

A few days after the Kickstarter appeal was posted, a slightly awkward young man in jeans and T-shirt stepped up to the microphone in an auditorium at the San Francisco Design Center. The crowd erupted in cheers. Mark Zuckerberg, the twenty-six-year-old founder of Facebook, was going to address software developers about changes that the site was going to make. On the surface, there was something endearingly unpolished in his apparent nervousness set against the gleaming graphics and stagecraft, like a high school play put on by venture capitalists. A focus on the bumbling style was a sure way to underestimate the breadth of the day.

"We think that what we have to show you today will be the most transformative thing that we've ever done for the web," Zuckerberg said.

That day, Zuckerberg and his chief technical officer, Brett Taylor, ran through a series of changes that would give Facebook a presence on other websites with "like" buttons. When clicked, these like buttons would alert the user's friends—and anyone else who viewed their Facebook pages—of their interests: what they had been looking at and, presumably, thinking about. The likes of users were a form of currency that Facebook would negotiate into advertising dollars. When other changes were added in, including the ability to watch what people were reading away from the Facebook site, what Zuckerberg and his backers were proposing was nothing less than becoming the tollbooth and gateway to the whole Internet.

By following the interest of its users across cyberspace, Facebook would plant its flag, and eye, on every corner of the web. What Facebook and Zuckerberg saw as a great expansion of human connectedness others saw as a silken trap from which there would be no escape.

At the same conference three years earlier, Zuckerberg had introduced the term "social graph" to describe the blueprint of connections between any two people. It was a variation on the Six Degrees of Kevin Bacon game, a dorm-room trivia classic in which the challenge was to link the actor to anyone else in Hollywood in six connections or less; the Bacon game itself descended from an early-twentieth-century notion that the expansion of communication and transportation had so shrunk the world that every person on the planet could be linked through a chain of acquaintances extending beyond no more than five others. In the early 2000s, a Harvard economics student applied the Bacon game principles to Harvard students; his sociogram, or social graph, was bared in Facebook friending.

In April 2010, Zuckerberg and Facebook were extending the social graph to record virtually any interest a person might express through the World Wide Web.

"The open graph puts people at the center of the web," he said.

The stunning implications were mostly unmentioned by Zuckerberg. If the social network functioned as an all-seeing vehicle for tracking people's curiosities, views, and interests, Facebook would control a vast database of human appetites. Its presumptive commercial value had already brought in billions in investment money.

With his friends, Zuckerberg had started the Facebook site when he was twenty years old, and in his view—apparently adopted after he had created his product; how worked out is the philosophy of any twenty-year-old?—individual life in the twenty-first century would be led in total transparency.

Every person arriving for the conference that morning was given a small RFID chip—the same kind of little radio frequency gizmos that Dan and Max had installed on the door of the computer office to send a tweet. Rather than tracking the movement of a door, the Facebook RFID chips would track people as they moved around the conference. When someone stopped, for instance, at a gaming booth, the chip would send a signal to a computer, and then to his Facebook page. The chips were in regular use for a "keg bot" by Facebook employees at headquarters: when people stopped at the office beer keg, a picture was snapped and loaded to an internal Facebook page devoted to life events of the keg, including when it needed replenishment. All good fun. But creepy.

The mass sharing would let a person going to a restaurant—one that had been rated on Yelp, the crowdsourced rating site—know immediately which of her friends had visited the restaurant and commented on it. Already "Yelp knows who I am, what I'm connected to, so they can give me a nice social experience," Taylor said. "What isn't possible is to take a part of this graph that Yelp has mapped out and connect to other parts of the graph. We want to have instantly social and personalized experiences everywhere users go."

The commercial power of such connections was not acknowledged. Nor was the limited control that a user had: if a friend on Facebook installed an application that wanted to vacuum up that person's acquaintances and their interests, there was little the user could do to stop it.

"This is a really significant step for Facebook. For years, we've been saying that Facebook is an open platform, but now for the first time, the likes and interests of my Facebook profile link to places that are not Facebook.com," Taylor said. "My identity is not just defined by things on Facebook, it's defined by things all over the web."

This may have seen an unalloyed, giant step forward to Zuckerberg, Taylor, and others in the bubble of Facebook's headquarters. Outside, the reaction was brutal. The toughest—because the nontech press relied on it to render the changes into nongeek-speak—was *Wired*, which ran a major essay that became the most-read story on its website for a month.

"Facebook has gone rogue, drunk on founder Mark Zuckerberg's dreams of world domination," wrote Ryan Singel. "It's time the rest of the web ecosystem recognizes this and works to replace it with something open and distributed."

The essay went on to recognize the important ground that Facebook had opened. But it argued that the company, and Zuckerberg, had not understood the importance of self-control.

"Clearly Facebook has taught us some lessons. We want easier ways to share photos, links and short updates with friends, family, co-workers and even, sometimes, the world. But that doesn't mean the company has earned the right to own and define our identities. It's time for the best of the tech community to find a way to let people control what and how they'd like to share."

Around that time, a professor at Columbia University studied Facebook users, asking their intentions and beliefs about their control of their privacy features on the site. Then he compared it with their actual settings. He found that 90 percent were sharing material that they had intended to keep private; a perfect 100 percent were not achieving their privacy goals.

The four guys met in the ACM room, door closed, and tried to figure out what had happened. They were just trying to scrape together enough money to write code for the summer. Now they were getting e-mails from Randy Komisar, a figure who Ilya reckoned was like the Steven Spielberg of Silicon Valley, which probably overstated Komisar's stature, but was near enough to the right order of magnitude for a kid who had been Dumpster diving not many months earlier. Komisar was at least a radio signal from a different galaxy. Dan was bemused: here they were, four guys sitting in a closet at NYU, and now important people were getting tapped on the shoulder about them. Rafi noted that they were in the last week of classes, and after that, they had finals. Max felt a twinge of guilt: he had not gotten very far on his senior paper. Going out to California the next day for a meeting was not in the cards.

At the very least, they agreed, they should call Komisar back, but be careful not to portray themselves as ready to roll out a gleaming new social network website in three months.

It took them another few days before they could organize time for a call to Komisar, who was in Menlo Park, California.

He was mildly startled when he found himself on the line with the whole team.

"Hi, Randy, I'm Max."

"I'm Dan."

"Hi, Randy. This is Rafi."

"Hey, this is Ilya."

Komisar was not coy.

"I love what you are doing," he said, by way of opening.

The guys wanted to turn up their cards immediately. For the next three or four months, they planned to build a code base, which they would make public and free for others to hack on. They were purposefully modest, as well as candid. One day they might want to do real business with Kleiner Perkins or Randy Komisar, and they did not want to burn their credibility with extravagant promises.

Komisar, though, wanted to talk about the project, and why he was interested. As mighty web fortunes were being built, largely on personal data that people had given away, he had become galvanized by how individuals managed their own information on the web. Property did not become private until it was owned. Taking the case of Facebook, which, depending on the day of the week, was supposed to be worth billions, Komisar explained his thinking.

"Let's say Facebook is worth $30 billion. That's $30 billion of my frickin' data. I just built a $30 billion business—or me and my 500 million friends did. That seems like a bad deal. We gave away our data too cheap."

For months, Komisar had been thinking about how to refactor the terms. He had spent time with the developers of Bynamite, who were building a little add-on to browsers that would, in effect, track the trackers—keep an inventory of the information that was being collected by advertisers as people surfed the web. It was a first step.

Describing his thinking later, Komisar said the goal was to "create privacy controls that effectively allowed for me to own my data, and thereby derive the value of that data. That doesn't mean I would keep it to myself. It means I would share it with who I wanted to, when I wanted to, for purposes that I wanted to. That was the context in which Diaspora came to my attention."

That was practically the language from their Kickstarter pitch. "With Diaspora, we are reclaiming our data, securing our social connections, and making it easy to share on your own terms," they had written.

In essence, web privacy could be seen as a commodity, not necessarily a right. A large part of the economy of the web rested on data mining; the question for Komisar was the matter of who got to trade in the privacy.

"What was great about the web—which was free information and content—couldn't exist in a world where privacy was strictly protected. If you strictly protect it, you would not be able to do effective marketing, and with no effective marketing dollars, none of that information content could be created and shared. Until someone owns that information, there's no interest, desire, to protect it. It's less a notion of blocking and more an issue of controlling, and thereby deriving value. If I could derive value from my data, I would suddenly become a lot more responsible with it. That became my angle, my thesis."

He asked about what they were going to do.

The current plan, they explained, was to go to Max's family's house in Lake Tahoe for the summer and code like crazy.

Komisar was charmed.

As they spoke with Komisar, their Kickstarter page had started to hum. Another thirty people pitched in that day. Word began to spread that there was a Facebook alternative. The Brooklyn hacker space NYC Resistor posted an item about the project. The news moved over to Reddit, a site where people shared links to news items that interested them. In Sweden, Peter Sunde, one of the founders of The Pirate Bay—an anticopyright, propiracy website that provided digital torrents of music, movies, and other material, and a legendary force among hackers—sent out a message on Twitter: "Looks like NYC. I don't know them, just love the idea. I know some people they know and they say they're cool!"

On Tuesday of that week, four senators called on the Federal Trade Commission to regulate online privacy. By the end of the first week, the four Diaspora guys had raised five thousand dollars, half of their goal, and were getting fifty to seventy donations every day.

A careful reader of the Kickstarter website could have seen another trail, which did not directly involve money. The Diaspora Four were showing that at a bare minimum, they had the temperamental chops to develop a big software project in public. Someone named Mike complained that they were giving updates through Twitter, a centralized service, when a perfectly good decentralized system called Identi.ca was available to perform the same function.

Rafi replied: "Good point, Mike!" and immediately posted a link to their brand-new Identi.ca account.

On the Kickstarter comments page, many people invoked a blog post written by Luis Villa, an intellectual property lawyer in California who had also worked as a programmer on many open-source and free-software projects. He said he supported Diaspora but was skeptical, and asked provocative questions, among them: Did they know about other projects that had tried similar things, and why did they think those hadn't worked out? Would they integrate any of the latticework of existing programs that would be useful in creating a social network website? There were more questions, and Max answered them all. In conclusion, he wrote: "The problem itself is not what server architecture, or what language we code in, or even what standards we support. While these questions are very important in making any solution better and more robust, they are all implementation details. The core problem is that users who want to share with their friends have no better alternative. We think any solution which addresses this problem is a step in the right direction.

"You said yourself that you have seen vaporware projects in this space in the past, and we have too. We think a project focused on getting something out there should be priority #1. We need to try a bunch of approaches, and see which ones take off, and we need to do it quickly.

"We want to be an independent code base because the four of us work fast and well as a team. Our arguments are short and solved by someone writing better code."

There wasn't a hint of defensiveness in the long, thoughtful answer.

In Max's closing words, he implicitly acknowledged that even other geeks who were philosophically aligned with them might have good-faith doubts.

"You say you don't have the time or money to work on such a project, one you yourself identify as important. That makes tons of sense to us. There are plenty of things we would like to do and we don't have the time for, and we are not even married or have jobs.

"If that is the case however, what would be un-pragmatic about giving four excited dudes who spent their last semester of school thinking about a problem you are 'worried-about-but-can't-deal-with-now,' twenty bucks so they can take an honest crack at solving it? :)"

Questions answered, the man pitched in.

A few days after they launched Kickstarter, Evan Korth set up a dinner for the four with Eben Moglen at Apple, a fancy restaurant in Greenwich Village. They told him the risks involved with Kickstarter, and Moglen said they should not worry.

"If you need a thousand dollars or so to get over the top, I'll get it for you," Moglen said.

They were planning to incorporate as a regular corporation; Moglen's arrangement with the Free Software Foundation was that he would not advise for-profit companies. Since they were not organizing themselves as a not-for-profit, he could not be their lawyer. But, he told them, if they ran into any problems, they could call on him. He just didn't want it to be widely known. The free-software world was highly ideological about such matters.

The recruiter welcomed Dan, hoped his flight from New York had been good, and explained that they were looking for promising young programmers.

"So," she said. "What kinds of things are you interested in?"

Dan thought for an instant about Diaspora, then about a technology product that had been released a few weeks earlier and had become an instant, startling sensation.

"The iPad is very cool," Dan said.

Its sleek style was galvanizing, of course, but Dan realized immediately that naming an Apple product was not the most politic statement. He was, after all, sitting in Redmond, Washington, at the headquarters of Microsoft, the prosperous but stodgy foil to Apple's hipness.

"Also, the Zune," Dan said, without conviction. Zune was a digital player that was Microsoft's attempt to compete with the iPod.

"I don't have anyone to meet you about them," the recruiter said. "We are interested in SharePoint."

Before he had gone into the office, Dan had shut off his phone, purposely. Now he could not wait to get outside. He did not care about Microsoft's SharePoint, a program that allowed people in an organization to develop and share websites. The meeting came to a quick end. He stepped into the hall and switched on his phone. What was the latest from Kickstarter?

When Dan got back to New York, there was another conference call from Randy Komisar. He had spoken with his partners. They had an invitation.

"You know what," Komisar said. "Why don't you come out here and spend the summer with us? We have an incubator right downstairs."

He stressed that there were no commitments by the other side.

"God, that's a great idea," Max said.

They were dizzy.

Every Wednesday afternoon, a group of invited academics, lawyers, and technologists gathered in a seminar room at New York University School of Law to discuss emerging issues in privacy, the Internet, and computers. It was convened by a law professor, Katherine Jo Strandburg, and Helen Nissenbaum, a professor of communications and computer science who was among the leading privacy researchers in the world.

Perhaps inevitably, Ilya Zhitomirskiy, the drop-in artist, turned up one afternoon. After the seminar ended, he blotted the anxiety from his palms and approached Nissenbaum. He told her about the Diaspora project. Nissenbaum, delighted, had one piece of advice: "Keep your eyes on the prize." Ilya left feeling commissioned by a great authority.

News about Diaspora ricocheted from family and friends, around the Internet and NYU. A few weeks later, it came back to the Nissenbaum seminar. Finn Brunton, a postdoctoral fellow, opened the session by giving the customary rundown of new developments. Among them was the Diaspora project, now raising funds on Kickstarter, and getting a strong response.

In fact, they had gotten to five thousand dollars, halfway to their goal, after a week. Five days later, they had hit ten thousand dollars. Then twenty thousand dollars. The money kept rolling in.

Among the people at the seminar when Finn Brunton mentioned the project was Cathy Dwyer, a professor at Pace University; she was one of the first academics to research privacy management tools in social networks, as well as commercial advertising tracking practices.

That night, she mentioned Diaspora to her husband, the About New York columnist at the *New York Times*—that would be me, the author of this book—and suggested the students would make a good subject for a column, pointing out that they had not only hit their goal of ten thousand dollars but were by then over twenty thousand dollars. I got in touch; they were insanely busy, but agreed to do a conference call on the evening of May 10. We spoke at eleven-thirty at night, which was the loin of the day, as far as they were concerned.

They had decided, ahead of time, not to tell me about their conversations with Kleiner Perkins Caufield Byers. The Kickstarter campaign was succeeding beyond their most exuberant hopes, and perhaps they really did not need to get involved with the venture capital world yet.

Ilya was not available, cramming for a final the next day. Rafi did not want to do the interview—arguing with Dan and Max that they had not written anything yet—but the others prevailed. Despite his reluctance, Rafi was the first to speak, explaining what inspired them.

"Eben Moglen gave a talk about how we're surrendering privacy to cloud systems," he said.

The value of the services, Max said, was negligible, given the scale of data collection that was going on. But they all used social networks.

"Certainly, as nerds, we have nowhere else to go," Max said.

Would you describe yourselves as nerds?

"Oh," Max said. "We're big nerds."

They were winningly modest and smart.

A picture session was organized for the following morning—the only time they'd be available, as it was the first of two graduation sessions for Dan and Max. All four gathered at the Courant Institute building, and posed in a classroom in front of a blackboard. Wearing a T-shirt and a watch cap, Dan sat at the front, holding open a laptop that showed the message they had posted on Kickstarter: "thank you friends." It also displayed the name of the project as "Diaspora*," the asterisk a graphical flourish.

As soon as the last shot was clicked, Dan and Max rushed off to meet their families. At midnight, my column was posted on the *Times* website:

"How angry is the world at Facebook for devouring every morsel of personal information we are willing to feed it?

"A few months back, four geeky college students, living on pizza in a computer lab downtown on Mercer Street, decided to build a social network that wouldn't force people to surrender their privacy to a big business. It would take three or four months to write the code, and they would need a few thousand dollars each to live on.

"They gave themselves 39 days to raise $10,000, using an online site, Kickstarter, that helps creative people find support.

"It turned out that just about all they had to do was whisper their plans.

"'We were shocked,' said one of the four, Dan Grippi, 21. 'For some strange reason, everyone just agreed with this whole privacy thing.'

"They announced their project on April 24. They reached their $10,000 goal in 12 days, and the money continues to come in: as of Tuesday afternoon, they had raised $23,676 from 739 backers. 'Maybe 2 or 3 percent of the money is from people we know,' said Max Salzberg, 22. . . .

"As they describe it, the Diaspora* software will let users set up their own personal servers, called seeds, create their own hubs and fully control the information they share. Mr. Sofaer says that centralized networks like Facebook are not necessary. 'In our real lives, we talk to each other,' he said. 'We don't need to hand our messages to a hub. What Facebook gives you as a user isn't all that hard to do. All the little games, the little walls, the little chat, aren't really rare things. The technology already exists.'

"The terms of the bargain people make with social networks—you swap personal information for convenient access to their sites—have been shifting, with the companies that operate the networks collecting ever more information about their users. That information can be sold to marketers. Some younger people are becoming more cautious about what they post. 'When you give up that data, you're giving it up forever,' Mr. Salzberg said. 'The value they give us is negligible in the scale of what they are doing, and what we are giving up is all of our privacy.' . . .

"'So many people think it needs to exist,' Mr. Salzberg said. 'We're making it because we want to use it.'"

# CHAPTER FIVE

NYU was holding the largest of its commencement ceremonies at Yankee Stadium, a forty-minute subway ride from Greenwich Village. It was May 12, 2010, a drizzly Wednesday in New York.

"No," Dan told Max. "I'm not putting on a dress to sit in the rain."

Max and his family rode the train, traveling with his aunt and uncle, Ann and Michael Wolf, who were going to host a party for Max that evening at their loft in the Tribeca section of Manhattan. Michael had a doctorate in computer science from Stanford, and although he had been out of the field for a decade, Max had been consulting with him regularly. Was he crazy to try doing Diaspora instead of going after a salaried job? Should they pitch for $10,000 on Kickstarter or $20,000? What about the interview with the *New York Times*? A good idea?

As it happened, the column on Diaspora ran on graduation day, and just before the Salzberg group boarded the subway at Canal Street, they checked the latest tally from Kickstarter. It was around $28,000, already up $4,000 that morning. That was a temporary end to their updates, as there was little cell phone coverage underground. About twenty minutes later, the train climbed to an elevated station. Kickstarter was at $32,000. When they reached the stadium, it was at $36,000. By the time graduation was winding up, they had hit $50,000. The donors were not just tapping in their credit card numbers, either. They were posting messages, too.

From "DMD" in Sydney:

Backed you for $30; make this happen :)cheers from Australia

From Thomas Klokosch:

I stumbled over your project while I was planning something similar myself. Now I'm an excited backer of your project instead. Cheers from Berlin.

Rather than go to graduation, Dan had gone to his mother's office in Times Square, in the same building where David Letterman taped his show. Dan commandeered her computer.

From Arvind Ravikumar, New Delhi:

I can see this will change the world! Just waiting to use it!

From his seat in Yankee Stadium, Max started tapping out a text to the other three, but one arrived preemptively from Dan.

"Holy moly!" he wrote. "I thought today was going to be big because we were graduating!!!"

Rafi and Ilya were watching in the computer club room, mesmerized, and vaguely terrified.

From Matt, San Francisco:

The rate of pledges is changing dramatically. The current hourly rate is better than $3,500 per hour.

Other people at Carolyn Grippi's office had heard about the project, and were tracking the changing total. Carolyn watched over Dan's shoulder with mounting concern. How much pressure could these four kids handle? Even giving the money back would be complicated.

From Richie in Louisville:

I have faith in you guys. Here's pasta money.

Rafi, especially, was becoming deeply troubled. He might have been the youngest of the group, but at age nineteen he had perhaps the most acute perspective on their capabilities and the scale of their ambition. The surge of donations was pushing past their modest hopes to have enough money for a summer of ramen and coding, toward something that they had not sought.

As the numbers soared above $50,000, Rafi, sensing that they had gotten in way too deep, texted the others: "Tell them to turn it off!"

But there was no way. Kickstarter rules would keep it up until the end of the month, as originally promised. The donors knew they had surpassed their goal several times over, but they were making a statement that went beyond funding a few kids.

From Communida Judia de Murcia, in southeast Spain:

We are the jewish community in murcia spain. we offer to translate free to spanish or hebrew.

Wrote "Woot":

I donated to this project because I want a social networking tool that is safe and gives me control of my information. I used to be a big proponent of Facebook, but they have sold us out. Even if this project fails, it proves one thing . . . users are fed up with Facebook's privacy trampling. Only good can come of this.

And a message from Mike Engray:

As a avid user of facebook, I wish the death of it. I am glad to see there is actually hope in people. Please keep this going. best of luck.

Another donor to the cause put the matter bluntly:

I do not care if the finished product works or not. I just want to see this ridiculous pile of cash sitting next to a giant banner saying FUCK FACEBOOK.

Spend it all on taxes, hookers, and blow, no worries.

Shortly after noon, while the graduation was still running, someone sent a tip into the website Geekosystem. Take note, the tipster said, of the code in the corner of the picture in the lead position on the home page of the *New York Times.*

Just over Ilya's right shoulder was a column of UNIX code including three incantations of a vaguely salacious abbreviation:

TOUCH

GREP

UNZIP

MOUNT

FSCK

FSCK

FSCK

UMOUNT

It was a version of the old UNIX code dirty joke having to do with mounting a file drive for some purpose, then inspecting it (File System Check=FSCK). Dan had written it on the blackboard a few months earlier, and somehow it had never gotten erased. The column was getting a lot of traffic on the *Times* website, moving onto the most e-mailed list, and it was being displayed prominently. Alerted that a risquéish image had slipped into the paper, the *Times* photo department cropped the online picture to remove the offending scrawl.

The guys were mortified but amused: they had simply gone into the first empty classroom to pose, and the photographer had taken a few dozen shots from all angles, including many that did not show the blackboard. When the page was being laid out, an editor picked one without noticing the ribald gibberish. It was a rip-roaring laugh along the geek telegraph wires—Max texted to friends, "Dan got four F-bombs in the Gray Lady"—but the Diaspora Four swore to all the friends hailing their derring-do that they hadn't realized what was written behind them on the blackboard. Anyway, pledges rolled in at a prodigious pace.

Through the afternoon, tides of friends washed in and out of the Wolfs' loft, joining parents and relatives. The Grippis came. All eyes were on

Kickstarter, or at least Max's were stealing looks at it: the family computer stayed lit, and every time the page refreshed, a new astounding total appeared. A runaway train had come to the party. Even Max's little cousin Sam, just past toddler, knew how to check the page. If the graduate and his friends strayed from the computer, Sam was sure to charge over with the latest report.

"Max!" he called. "You got another thousand dollars!"

Rafi and Ilya turned up, not only to congratulate their friends on their graduation but to have company as the totals swelled and their modest plans for three months of ramen were being torn up. Whatever notions Max and Dan might have harbored about a project that might somehow continue once the ten thousand dollars ran out, for Rafi, Diaspora had been strictly a summer project. He had not committed to anything beyond that. The waves of money that crashed against the Kickstarter site were a kind of chaotic force that he could not easily analyze, nor was he sure how to harness. They really hadn't done a thing yet.

Max broke into the reverie of his dismay with an announcement: "Guys, there's an e-mail from Randy Komisar."

What to do about the venture capitalists? Max's uncle Michael Wolf knew the stature of Kleiner Perkins in Silicon Valley, but he also thought that to build the kind of project they had in mind, they would need help—free labor and brainpower—from a community of software developers. If venture funding was behind Diaspora, why should volunteers contribute their sweat?

Komisar was just checking in to pass along a message that he had received that morning from another person at his firm, and Max read it out loud: "'Are these the guys you were telling me about at the partners' meeting?'"

The message, Max announced, had been sent to Komisar by none other than Bill Joy.

A name not widely known outside technology fields, Joy was legendary as an author of software code that was a cornerstone of all modern computing. He also had been an important shaper of the protocols that allowed networking. *Fortune* magazine had called him the Edison of the Internet.

To such circumstances, the reserved detachment of even the coolest, most detached of the group had to yield. Rafi threw his arms around Dan and hugged him.

One of Max's closest friends, a high school classmate named Dan Goldenberg, had come from Boston for graduation. Early that evening, he was on a bus back. He texted Max: "The bus has wireless, and the girl sitting near me has her laptop open to your Kickstarter page."

By the end of the night, they were over $100,000 in contributions, and they had another three weeks to go in their Kickstarter fund-raising period.

When it was all over, they had been given $200,642 by 6,479 people from around the world. Another 40,000 signed up to follow them on Twitter.

Of course, there were skeptics, and among the earliest to surface was Matt Rogers, writing on an influential tech blog, *Download Squad*. Amid some trivia—he complained that the Diaspora dudes were so immature that their video was terrible—he made some strong, loud points. Membership in Facebook was not compulsory. Its growth reflected either consent or indifference on a mass scale to the network's predations. In other words, the people were voting with their clicks. Like the first-chair violinist playing an A note as an orchestra tunes up, Rogers's contempt was a tone adopted by much of the criticism that followed. He wrote: "The bottom line here is that so few people will ever use this vaporware (were it ever to materialize in the first place) that it simply won't succeed. It doesn't stand a chance against a network like Facebook—no matter how evil it may be—because Facebook isn't evil by accident. Its massive numbers of users allow it to be that way. If the majority of people actually cared enough about their online privacy, then they'd leave Facebook. They don't, so it's difficult to imagine that these kids will ever see their little cloned project become a network of 10,000 users, much less one that includes 'every man, woman, and child' like they actually had the audacity to state as their end-of-summer goal."

He continued: "Basically, if you donated money to these guys, you didn't participate in some grand assault against Facebook's foothold on the Internet. You probably paid for an appletini or two (of thousands) that will be consumed over the course of the next few summer months . . . as they party their faces off. And why shouldn't they? They only asked for $10,000 in pledges. They've just pulled off a heist worthy of a bad Hollywood movie."

Yet among the donors to Diaspora was a savvy fellow traveler: Mark Zuckerberg, the founder of Facebook. He pledged one thousand dollars. When Facebook was getting started, it could not support a giant server to hold all the pictures that people wanted to share. So Zuckerberg and his friends had tried a decentralized approach for photographs—having people store them on their own computers, which would be designated as servers with a piece of software provided by Firefox called Wirehog. (Facebook's president, Sean Parker, who had been with Napster when it got into legal trouble for peer-to-peer music sharing, said Wirehog would get into the same problems, and killed it because he deemed it "illegal.")

In an interview with *Wired* on plans to improve privacy controls in Facebook, Zuckerberg said he admired the Diaspora guys for exploring the possibilities of decentralization.

"I think it is cool people are trying to do it," Zuckerberg said. "I see a little of myself in them. It's just their approach that the world could be better and saying, 'We should try to do it.'"

And in another article, *Wired* noted that Diaspora's goals were shared by other projects, mentioning Appleseed and OneSocialWeb, and tracked down Eben Moglen. He wasn't picking favorites, he said, but believed that the reception to Diaspora's plans was an important moment for the free-software movement.

"The funding is not through the capital market and not through the venture capital system," he said, "but through civil society. This is a signal to Facebook, and I am sure Facebook is getting it."

# CHAPTER SIX

I n twenty-first-century terms, the blacksmith and goldsmith Johannes Gutenberg, born around 1395, was pure hacker: he drew from a palette of technologies, rag paper, metals mined near his town, and oil-based inks, then synthesized them with a machine for making type move. Thus, printed books. Months after Martin Luther nailed his "Ninety-five Theses" to the door of the church in Wittenberg, some 300,000 copies of the tract were distributed across Europe and Germany, a development unthinkable before the printing press. Democratic revolutions followed its adoption, as did representative democracy; so, too, did decades of religious war. As a force for change, few events in history could approach the invention of the printing press. The development of the Internet surely comes closest.

The surface area of digital communications has grown by the day, grafted onto the skin of the planet. From the richness of the World Wide Web grew an information economy, and then an ecosystem of surveillance. Much web development was driven by commercial incentives: build useful platforms like social networks, or restaurant recommendation sites, so you can find out what people were saying and showing to one another. Then sell marketers on a chance to form a perfect union between appetites revealed and purveyors of things to satisfy those hungers. That was only part of it. Those seeking to control or monitor communication for political purposes also gave hot pursuit.

The funding of Diaspora was but one of the nearly uncountable simultaneous wrestling matches going on around the world for control of cyberspace. The year that Diaspora roared into view was a season of overture, of gathering disruption in countries that had been lodged in cultural and political permafrosts. The same Internet that had commissioned the four NYU students to save the world from Facebook was also pulling down the pants of despots and strongmen in swaths of the Arab world. In fact, it was using Facebook to do so. In Egypt that summer, a page on Facebook, Kullena Khaled Said, was drawing hundreds of thousands of visitors. It commemorated a young man who died from a severe beating while in police custody. The administrator had disguised his location by sending his communications through routers in many countries, deploying Tor, a tool specifically created to circumvent surveillance. Tor stands for The Onion Router, because the multiple layers of relays were the baffling element that hid the identity of the user.

Said an Egyptian activist, Esraa Rashid: "No one from the opposition or the government knows who started the page."

Neither, for that matter, did anyone from Facebook. In fact, that anonymity violated Facebook's terms of service, which does not permit fake accounts. After the page had been up for months, and just as it emerged as a trumpet summoning crowds to Tahrir Square for protests, Facebook unilaterally shut down the page. The administrator had to be a real person. Rules were rules. The group scrambled and got an administrator who was beyond the reach of the Egyptian government. The page was restored.

That episode provided a glimpse of how Facebook, and the other transnational giant, Google, had become the global equivalent of shopping malls in the United States. They were territories that had become laws unto themselves: banning or restricting political protests, demonstrations, and solicitations as they—and they alone—saw fit, employing sweeping approaches that could not be used on the public streets because of free-speech protections.

A crusading blogger, Wael Abbas, had been posting videos of torture by authorities since at least 2006, and one of them, involving the sodomizing of a bus driver who refused to pay a bribe, led to the prosecution and conviction of three police officers. In 2007, Abbas's YouTube account was shut down and more than one hundred videos removed; later, YouTube, which is owned by Google, restored the account and said that it had acted because sufficient context for the videos had not been provided. (A video of a cat stepping into a bowl of milk was self-explanatory; apparently not so for videos of handcuffed men being beaten by men in uniform.)

In countries all over North Africa, the ground was heaving as the trapped forces of candid, vigorous speech suddenly found in the Internet a seam in the geological formation of their societies.

Having grown up in the sclerotic civic life of Yemen, where the same man had been president, as of 2010, for thirty-two years, Walid Al-Saqaf, an energetic computer scientist, took up journalism graduate studies in Sweden. He created a web page of uncensored news about Yemen.

"Day and night you listen to propaganda from government sources," said Al-Saqaf. "The only sources with some breathing space, free space, [are] online."

Back in Yemen, he awoke one Saturday morning and checked into the site. But nothing loaded. The browser hung in the suspended animation of "File Not Found" messages, no matter how many times he tried.

Tunisia, which liked to call itself the flower of North Africa, was proud of its thriving modern economy, stability, and relatively progressive posture on social issues. Women had rights that were the equal of those enjoyed in Western countries.

Yet since gaining independence in 1957, it had known only two presidents; the incumbent, Zine El Abidine Ben Ali, had taken power in 1986, when his predecessor had been declared unfit for office. "In America, people cannot name their presidents. In Tunisia, during your lifetime, you know one. Or 1.5," an activist, Houeïda Anouar, said. "A guy in one of the opposition political parties said in his whole life he had two minutes on TV."

During a hunger strike by people calling for reforms, Anouar spoke every day with the doctor who was monitoring their health, made videos, and then would upload them to a blog. One day, the blog had been hacked: the figure of a man in black and white with a red hat had been inserted onto the page. Anticipating such an attack, she had backed up the blog, and simply replaced the damaged pages. Three times she had to restore the blog from a backup, Anouar said. "Then they blocked access to the website," she said. "I had mirrors outside the country"—that is, her website existed in duplicate on servers in places where Tunisian authorities had no power.

"That was when they blocked the Internet to my house," she said.

Censorship was becoming a global norm. More than forty countries were filtering Internet content in September 2010, nearly double the number that were doing it in 2007. "Saudi Arabia held back on widespread deployment of the Internet until they could install a filtering system for their country," said Rob Faris, the director of the Berkman Center at Harvard University, a think tank that specialized in digital freedom. "Iran was an early adopter. China produced its own solution."

With different regimes of monitoring, surveillance, and censorship, the borderless regions of cyberspace were being renationalized. The faucets and valves could be selectively turned off.

When Al-Saqaf's website went into suspended animation, it turned out that Yemeni officials had "simply clicked a button on software they use to filter content in Yemen," Al-Saqaf said.

The filtering software was provided to Yemen by a company called Websense, based in San Diego; eight other Middle Eastern and North African countries used filtering systems provided by companies in North America. The software can be used to block objectionable content in schools or libraries, keeping out pornography and malware; it also makes it possible for authoritarian regimes to knock out websites like the one run by Al-Saqaf. Each brand of filter maintains a database of banned sites, a blacklist of forbidden content.

"Often pitched in the first instance for use by parents, schools, and workplaces, these technologies can also be sold to make filtering easy for entire countries: Once the underlying infrastructure is set up, the censors need only activate the tool and select the categories they wish to censor," a report by the OpenNet Initiative found. Websense stopped updating its database of blacklisted sites for Yemen when the censorship was made public, and said it was a violation of the terms of service to use it to suppress speech. Nevertheless, the genie was out of the bottle.

Al-Saqaf's immediate concern was not who or what had made the shutdown of his website possible but how it might be reversed. He organized an anticensorship campaign, then realized that he needed a technical solution. With others, he developed a system of chutes and ladders that routed web traffic through multiple servers. That would bring the user to his unfiltered server. This, however, was a risky approach: his servers would buckle if everyone who wanted to play online games, or run porn videos, tried to stream the content they wanted through his server.

So Al-Saqaf executed a move of digital jujitsu. He took the basic design of filtering systems—the blacklist—and turned it inside out. His filter would be a white list of essential sites, approved by him and a few trusted allies. Sites on the white list could pass through his network of servers designed to obfuscate the traffic; those not on the list would pass through normal channels, without the layers of disguise. He called it split-tunneling, and it would conserve bandwidth, a precious resource. Video games would be out. News portals, opinion sites, social networks would be in.

He called the site Al Kasit, Arabic for "The Circumventor." It spread quickly. By the fall of 2010, some twenty thousand URLs for sites from all over the world were on his white list. Al Kasit became a tool for news consumers in Yemen and in sixty other countries, including many where people felt that impolite, impolitic speech was being monitored or blocked. Among these were near neighbors to Yemen: the United Arab Emirates, Jordan, Sudan, Bahrain, Syria, Kyrgyzstan, China, Libya, Tunisia, Egypt, Morocco, Turkey, Hong Kong, Oman, India, Qatar, Iran, and Saudi Arabia.

"There is a state of denial in Yemen that there is censorship," Al-Saqaf said. Even so, he was called to a meeting by an adviser to President Saleh, who asked him to shut down Al Kasit.

Al-Saqaf was prepared. "You've done what you can do, I've done what I can do," he said. "Here's a gift."

He handed the president's adviser a CD that would set him up on Al Kasit.

"You can use it to access censored sites," Al-Saqaf said.

The strongest hands played against an unrestricted Internet are not simply government regimes with a history of intolerance for dissent; the U.S. National Security Agency has tried to unravel Tor, The Onion Router, one of the most powerful circumvention tools for activists like the administrator of the Facebook page for the Egyptian opposition. It is a classic case of one hand undermining others: Tor was created in 2002 with early funding from the U.S. Naval Research Laboratory, which had an interest in secure communications. Since the success of the project depends, in large part, on a global network of nodes, it continues to receive about 60 percent of its funding from the U.S. government. The purpose is to give practical muscle to the "soft diplomacy" initiatives of the Obama administration's Department of State in human rights advocacy. The spy agency's logic was that whatever useful camouflage Tor provided to dissidents, it also served as a hiding place for forces who were a danger to society. "Very naughty people use Tor," an internal NSA presentation declared. The code name for the NSA operation to crack Tor was "EgotisticalGiraffe," the *Guardian* newspaper reported, basing its account on documents released in 2013 by former NSA contractor Edward Snowden.

In 2010, as the Diaspora project was getting started, Google released its first "transparency report" about requests from government and others to remove content from parts of the Internet under Google's control. It announced in November 2013 that the requests from governments for information about users had increased by 100 percent in the three years since the first report. The United States led all countries by a wide margin, seeking data on nearly eleven thousand people, and an uncertain number of requests under the Foreign Intelligence Surveillance Act. Google said it was bringing a suit against the United States because the Department of Justice contended that the company was not permitted to disclose those requests.

When the 2009 election in Iran returned Mahmoud Ahmadinejad to power in a supposed landslide, Isa Saharkhiz, a fifty-seven-year-old journalist, became a conduit between reformers and foreign reporters. "After the disputed election, they came and raided the house and took the computer," Mehdi Saharkhiz, his son, said.

His father hid out in Tirkadeh, a small village in northern Iran. "He was in full disguise there," Mehdi said. "He kept his phone off all the time, except when he was making calls.

Nevertheless, intelligence agents hunted him down. "They told him that they tracked him through the cell phone," Mehdi said, "the few minutes a day he was on."

The agents had a powerful tool: for several years, Nokia Siemens Networks had been supplying the two leading mobile network providers in Iran with items from its "Intelligence Solutions" catalog, which delivered "tailored solutions for lawful interception, monitoring and intelligence analysis to law enforcement agencies, government agencies and authorized groups worldwide." The brains of its product line, purchased by the Iranian companies, were monitoring centers trumpeted by Nokia Siemens for the power of their surveillance capacities: "Its unique modular Front-End and Back-End architecture allows the monitoring and interception of all types of voice and data communication in all networks, i.e. fixed, mobile, Next Generation Network (NGN) and the internet."

Using a panther for a logo, Nokia Siemens said the products were built for pulling together scattered bits of information and seeing where they led. "It is the smart analytical tool for intelligence personnel and analysts, enabling them to trace the track and helping decision-makers to fulfill their mandates: To identify and predict trends, patterns or problem areas that require action."

The extent of this power was reported by Hanna Nikkanen in *Voima*, a Finnish newspaper, in an article that quoted from a manual for the equipment: "Collecting interception data is a process which takes place in the 'background,' assuring that the intercepted target (end user) is never aware of a possible interception. . . . The maximum number of simultaneous active interception sessions is 50,000."

During his interrogation, Isa Saharkhiz was beaten and injured. He was sentenced to three years in prison for actions deemed subversive and disrespectful to the supreme leader, and his health declined. "We understand that he is now in a wheelchair," Mehdi said.

Nokia Siemens, having sold its surveillance gear as the avant-garde of technology for governments of all stripes, was called before the European Parliament in June 2010 to explain its provision of such equipment to repressive regimes. As it happened, European regulators had themselves been at the forefront of pushing for the integration of more powerful surveillance and tracking standards in mobile phone networks.

"Governments in almost all nations required operators to deploy Lawful Interception as a condition of their license to operate," Barry French, a Nokia Siemens executive, told the hearing. This was a passive component, he said, that most likely was already in the phones used by everyone in the hearing room. "And for good reason: to support law enforcement in combating things like child pornography, drug trafficking and terrorism."

The monitoring centers were a separate, but necessary, component from the Lawful Interception features, he said; they gave enormous power to law enforcement agencies, at times bypassing oversight functions.

"Monitoring centers are, in our view, more problematic and have a risk of raising issues related to human rights that we are not adequately suited to address," French testified.

So Nokia Siemens had started to get out of the monitoring business a few months before the disputed 2009 presidential elections in Iran, he said. But not before it had installed a system that made it possible for the tracking of Isa Saharkhiz and others.

"We believe we should have understood the issues in Iran better in advance and addressed them more proactively," French testified. "There have been credible reports from Iran that telecommunications monitoring has been used as a tool to suppress dissent and freedom of speech."

Nokia Siemens could not reinvent history, he said, but it could learn from it. And he reminded the legislators that there were tensions among ideals and technology, and international agreements on standards.

"Consider the fact that the systems that we provided to Iran were designed to implement a right that the [International Telecommunications Union] has said is held by member states, and are required by law in the vast majority of those member states," French testified.

Societies evolve, he said, leaders change, and governments do not remain static over the life of a set of technologies.

"We are always at risk of finding that we have deployed technology that seemed appropriate for use by one government only to find it misused by the next," he said. A fair enough point in the abstract, but the Iranian regime was already authoritarian in character when Siemens Nokia supplied the surveillance equipment.

An American company called Blue Coat Systems, of Sunnyvale, California, provided surveillance and censorship equipment to more than a dozen countries, including ones with histories of vivid abuse of human rights—like Syria, Saudi Arabia, and China. Other countries included Russia, Bahrain, and Thailand, according to a report in 2013 by the Citizen Lab Internet research group at the University of Toronto.

It was a matter of definitions. One person might see a surveillance package as a sentinel for people haunting cyberspace with disfigured appetites for children, or notions of how to slaughter civilians; another might see it as a way of hunting and crushing dissent.

As the four Diaspora guys set out to build their project, the Internet was still in its big bang moment, the clouds of its atoms nowhere near settled into recognizable forms. Its short history was summed up by Rob Faris of Harvard's Berkman Center during a conference in Budapest on digital liberties: "A bunch of smart people invented the Internet; another person added World Wide Web. Soon tons of people were saying, hey, this is great. We can share recipes for cookies; can share tips for gardening and knitting; we can create groups to share science fiction."

It was also, he pointed out, a source of recipes for bombs, a platform for incitement and subversion and persecution, and, sometimes, a tool of liberation. And it was grafted into every dark corner of the human psyche.

"Millions of other people said, hey, the Internet is great for porn as well. Others," Faris said, "offered up ways to more effectively commit suicide."

# CHAPTER SEVEN

**M**ax turned his head, unable to watch. Out of the blue, Ilya had just volunteered to show a bit of Diaspora's progress to a group of about forty fellow hackers—people who, for much longer, without any attention from mainstream news organizations like the *New York Times,* and certainly with no monsoons of Kickstarter cash, had been working toward essentially the same goals as the four young men from NYU. In mid-July, the group had gathered in Portland, Oregon, for a summit.

Just when the serious business of the day was winding down and people were packing up to go sample the city's artisanal beers, Ilya volunteered to show one of the group's gurus, Evan Prodromou, how a particular Diaspora feature worked.

"This is not the moment to do a demo that we've never actually tested," Max muttered. "And not in front of this audience."

But Ilya already had his laptop open, and Prodromou was leaning over his shoulder.

"This is when things never work," an onlooker commiserated with Max.

Another person, in the back of the room, Jon Phillips, was blunt about Diaspora. "These guys are vaporware," he declared. Prodromou was also a skeptic. One purpose of the summit, he had said, was to dispel the notion that some "messianic" force by itself could achieve their goals. But if people there hadn't figured out by then that the Diaspora Four had no illusions about being messiahs, then Ilya's wide-eyed guilelessness, his unguarded exuberance about sharing an early draft of the work in progress went a long way to allaying such fears.

Long before Diaspora lit up Kickstarter, Prodromou and other members of the group in the room had toiled in anonymity to make it possible for people to use the web as a social tool, on their own terms. Share a photo? Give a thumbs-up for someone's comment on an article that had been posted? Track down old friends? All these things could be done without having to go through Facebook or anything like it. With the right tools, they believed, each person could and should be sovereign over what he or she chose to share. Moreover, there was no reason that a person on Facebook should not be able to connect with a person in another network.

Those were the general principles of what this group of hackers called the federated social web. Services like Facebook fenced in their users, deciding what sites they could interact with and how they could display information. The better model, in the view of the federated social web group, was the telephone: no matter what brands or models of phone two people were using to converse, regardless of whether they were customers of AT&T or Verizon, they could call each other and send text messages. At the summit, the hackers spoke about ways to make that happen on the web.

At its essence, the World Wide Web created by Tim Berners-Lee and his generation was entirely federated: it consisted of databases of documents that were stored around the world in millions of digital libraries, called servers, and connected through the Internet. Regardless of who had created them or how they were stored, web browsers made it possible to view any of them that were not restricted by passwords. There was no high wall to block users from communicating with someone else on the web just because the two parties used different servers, for instance, or one person had text files while another had photos. Similarly, e-mail is federated: a message can be sent from one service to another, like a person with a Gmail account writing to someone with a Yahoo! account, or a personal mail server. The hackers at the federated social web summit saw no reason those principles should not apply to all kinds of personal sharing.

"From the point of view of a typical social website, if you don't have an account on that site, you don't exist," Prodromou said. "The only way for your friends on that site to interact with you is if they invite you to join the site. Despite the fact that there are hundreds of other social networking sites on the web, almost every single one works as if there were zero other social networks on the web."

To get around this, Tantek Çelik, another of the organizers, explained that loosely affiliated hackers had spent years developing and agreeing on short strings of code, microformats. Each string performed what seemed like an elementary task. For instance, one allowed a user to authentically identify himself to another, and made it possible for the receiving party to answer back with a persuasive confirmation of her identity. Another let a single message be published to a list of contacts. They might have seemed simple, but it was essential that people making software reach a consensus that these strings—rather than some other formula—were the methods that would be used. By the agreement of this community of makers, some of them in start-ups, others in major corporations, these microformats were the building blocks of a social web. They belonged to no one. It was principle-driven work. Plus it was practical: why shouldn't they be able to communicate to anyone, anywhere?

"The only way we are going to move social networks forward is if we can interact with our friends across social networks," said Chris Messina, a Carnegie Mellon graduate who also had worked on the microformats before joining Google.

These microformats were wires, pipes, and valves, not exciting, but essential infrastructure. The Diaspora group had not said much about using them in building their project, and their arrival in this crusade, with their $200,000 in crowdsourced money, to the accompaniment of trumpets by the *New York Times,* was the occasion of grumbling, encouragement, and worry. Blaine Cook, at thirty, was one of the elders. Working as a lead developer for Twitter, Cook had created a microformat that allowed a Twitter user to connect to other applications without revealing password and identification information. After leaving Twitter, he continued to fiddle with the standards. He had written to Diaspora and was eager to meet them in Portland. It was a relief; he was happy to see them succeed, but not to abandon the work that had already been knitted together.

"When I first saw the project, they had their own technology stack ideas," Cook said. "All of us were like, 'Oh, God, they'll set us back two years with the technology that they were proposing.' It was massively complicated. It was going to be the sort of thing where people look at it, and it's like, 'Oh, this technology doesn't work, and let's abandon this idea.'"

Max and Ilya assured him that Diaspora was happy to discover a community of fellow travelers, and fully intended to use their protocols. "We're very glad that you guys have done all this already," Max said.

Max turned to Dan. "Making standards is a thankless, awful job. It's like giving a root canal while you're getting one," he said.

"Did you just think of that?" Dan said.

In the front of the room, Ilya had his laptop open and was showing Prodromou how Diaspora could use one of the micoformats to communicate with StatusNet, a microblogging company run by Prodromou.

"I was pleasantly surprised," Prodromou wrote later. "Hard to believe how quickly this is moving along."

Tantek Çelik said that all of their work had to be proven in real life, beginning with themselves. It was a cardinal principle of development: use the software yourself. Dunk it in a bath of real life. If you're making dog food, eat it.

"As soon as I build something, I try to dog-food it immediately," Çelik said.

Interest in federation was hardly limited to the group in Portland; Diaspora had shown that there was excitement around the world. Not only could it mean better tools for socializing; it could also be a place for more vibrant political speech.

"Consider that informed citizens worldwide are using online social networking tools to share vital information about how to improve their communities and their governments—and that in some places, the consequences if you're discovered to be doing so are arrest, torture or imprisonment," wrote Richard Esguerra of the Electronic Frontier Foundation, the influential force on digital and human rights issues. "With more user control, diversity, and innovation, individuals speaking out under oppressive governments could conduct activism on social networking sites while also having a choice of services and providers that may be better equipped to protect their security and anonymity."

# CHAPTER EIGHT

few minutes after nine, Ilya dropped his bag at the desk, grabbed a plate, and, working from a groaning board that he could have navigated with his eyes shut, speared waffles and bacon. Dan was already parked at a table, sipping coffee. Max was deep in conversation, his plate cleaned. Rafi was a few steps behind Ilya at the buffet. All armies, including coders, move on their stomachs.

By that Friday morning in early August 2010, the four guys had spent two months holed up in a big software development laboratory in San Francisco, working anonymously and ferociously. Before they left New York, everyone wanted a piece of them. The Grippi family discovered a television news crew parked outside their home one morning. Reporters from all over the world were hunting down any of the Diaspora guys to hear nerds declare war on other nerds. Anyone connected with Diaspora was hounded by a press corps eager to speak with the "Facebook Killers"—a title that was the snickering figment of a headline writer's mind, embodying an ambition the four guys had never uttered, or, as far as could be told, entertained.

"We have to just go and code," Rafi had said. There was no point spending huge amounts of time to correct a twisted version of their hopes. They would otherwise never get anything done.

Other than to post a grateful blog item that was distributed to their Kickstarter list, saying that they were going to put their heads down and work for the rest of the summer, the Diaspora Four said nothing about where they planned to do it or when they would emerge. They vanished. The windfall coinage of their celebrity in public had stunned them. Every second, eight new people were joining Facebook. As Goliath swelled, so did the importance of the Davids.

They were invited to chat with a man named Mitch Kapor, an elder and eminence of the tech world whose philanthropies, businesses, and causes occupied a few floors in an old warehouse building south of Market Street in San Francisco. People brought their dogs to work. The guys were charmed.

"I love what you're doing," Kapor told them. "I want you to win."

In 1982, Kapor cofounded Lotus Development Corporation, which offered a spreadsheet, database manager, and graphics package called Lotus 1-2-3. The company's business plan called for $1 million in sales the first year. It sold $53 million. Two years later, the revenue was $153 million. In the rearview mirror of history, the reason is obvious. Lotus showed how numbers move. It mapped patterns of change and rate. Trends buried in the density of numerals now were visible, backlit by Lotus. A simple spreadsheet program was an epochal change for people who labored with pencil, paper, and calculator to figure or forecast profit and loss, rate of return. Businesspeople, academics, and scientists lined up to buy computers so they could run Lotus. It became the first killer app. Bill Gates had once predicted that there would be a personal computer on every desk. Lotus showed why.

After selling the business, Kapor dove into digital progressivism, cofounding the Electronic Frontier Foundation and playing an obstetrical role in the birth of Firefox, the open-source browser that had changed the Internet. He had become convinced that Mark Zuckerberg was vastly overreaching and that an alternative to Facebook was needed. Earlier that year, as Kapor was reading an article online, a notice popped up informing him that an acquaintance had read and "liked" the same article. Furious, he picked through the setttings of his own Facebook account, hoping to minimize the appropriation of his interests.

So he was ready to hear about the Diaspora project. They could check in and keep him up to date, Kapor told them, and he would think about ways he might be helpful. In a pleasant haze, the guys wandered back to their project, his words ringing in their ears. "Keep your heads down, keep working," Kapor had told them.

Venture capital firms were writing. Besides the federated social web gathering in Portland, they had—thanks to Mitch Kapor—an invitation to speak at Mozilla, creator of the web browser Firefox, and at Razorfish, an online advertising firm. Bill Joy—the Edison of the Internet—wanted to meet, a scouting mission for the venture capital firm of Kleiner Perkins Caufield Byers. A writer with *New York* magazine who was friendly with their mentor, Evan Korth, had persuaded them to break their self-imposed embargo for a profile to be published in the fall.

The first big decision was what to do about the open offer from Randy Komisar to set themselves up for the summer in the incubator space at Kleiner Perkins on Sand Hill Road, the boulevard of gold in Silicon Valley. His invitation had arrived before the waves of publicity and the cosmically fortuitous timing of Facebook pissing off much of the planet. Not only was the notion of being in a VC incubator beyond any expectation, it was beyond conception. They had tentatively accepted. Yet that did not mean they could drop their Kickstarter pitch or the "four dudes eating pizza" tableau. Whether they were working in the basement of Max's mom's house or at the bottom of the Kleiner Perkins office, they would still need ramen money.

The sluice of donations changed everything, making it emotionally implausible for them to move in with venture capitalists. "People expect that we're going to go live in a shack somewhere and do this," Dan said.

Ilya had been charmed by Randy—they were kindred enthusiasts—but he felt a tighter commitment to the people who had funded them to keep a kind of pure spirit of what they were doing. "We can't go the venture capital route now," Ilya said.

Max and Rafi agreed. Moreover, they would not have to suffer for their principled stance. Another opportunity had emerged. Rafi's brother Mike, an experienced programmer, had just started work at Pivotal Labs, a leading software laboratory in downtown San Francisco that developed programs for companies like Best Buy and Twitter. Pivotal was a master workshop for "agile development," which encouraged programmers to be adaptive in their solutions, a departure from the more rigid approach of working off a master blueprint. It billed the services of its programmers at about fifteen hundred dollars a day. Start-ups also hired Pivotal to show them the tricks of the trade, and would take up residence in the Pivotal offices while being nurtured. After the engagements finished, some of those start-ups ended up staying in the Pivotal space to keep working.

Diaspora certainly wasn't going to hire Pivotal, but Rob Mee, the company president, thought they were onto a good idea, if quixotic. He admired them. Plenty of people talked about building that kind of alternative social network, but no one had done it. Moreover, the guys were working in a programming framework called Ruby on Rails, a relatively new approach that was one of Pivotal's basic tools. So Pivotal happened to have the world's leading assemblages of experts in the very language they were using. Mee invited them to come work in the space for free.

With that offer in hand, they phoned Randy Komisar.

"It's really important to the spirit of the $200,000 that we raised that we be independent, and that we be perceived as fully independent," Ilya said. "We're concerned that the people who have given us this money want us to be countercultural. We are concerned that being in your incubator would be an affront to that."

"You know what?" Komisar said. "I buy that. So let's get together when you get out here. I'll do whatever I can to help you. Feel free to call on me."

By early June, they had set up a four-man code factory in Pivotal's office on lower Market Street in San Francisco.

Spread across a wide-open floor, the Pivotal developers could count on a full breakfast every day, lunch on Wednesdays; a row of bikes hung from a wall; during lunch hour, a court of Ping-Pong tables next to the dining area was heavily used.

That was a casual gloss on a company that had a precise, disciplined approach to software development, the antithesis of the hacker ethic of staying up all night, writing code, and shoveling down pizza. The Pivotal programmers—they called themselves Pivots—worked from nine to six, and almost always in pairs, two people seeing the same screens as they worked. Pair programming helped to prevent white-line fever, the paradoxical effect named by long-distance truckers who look at something for so long that they no longer can see it. In programming, the second set of eyes, Pivotal had found, cut down on both piddling errors and major logical flaws. It also built mentoring into the work. The programmers changed partners every day, avoiding the ruts that might develop between two people; it also ensured that the entire team was fluent in all elements of the project. The structure had been a revelation to Max, Dan, Ilya, and Rafi, who were accustomed to swarming over a project and keeping the long night hours of college students on deadline benders, with no fence between daylight and night.

On their first day in the office, absorbing this new culture, Rafi's brother Mike came over to chat with Max.

"Congratulations," he said. "At the end of the summer, or when the project is finished, you'll be able to get jobs as software engineers."

That was not where Max saw himself going.

"I want to be the CEO of a start-up," he said.

Surrounded by developers whose programming talents were worth millions of dollars a year, the four NYU guys burrowed into what was practically the perfect cave—and near-perfect anonymity.

Nearly.

One afternoon, Rafi came up the stairs from the subway near the office on Market Street, and strolled blithely to the Pivotal offices.

A moment later, a stranger sent a tweet out into cyberspace:

I think I just got off the subway after Raphael from #Diaspora!

It initially unsettled him, but the others were simply amused.

For the first time in their lives, Max, Dan, Ilya, and Rafi were not working to someone else's schedule: no academic calendar, no summer job, no expectations. They were on their own. The funds from Kickstarter had been deposited into a bank account, and Dan's father, who had taken early retirement from Citibank, handled their finances. Their budget was austere: one-thousand-dollar monthly stipends, plus a housing allowance. Additional expenses were decided on an ad hoc basis, and Dan, the de facto treasurer, was notoriously careful with the money.

One day during the summer, Randy Komisar dropped by to chat with the group. Mike Sofaer walked into the session with Komisar. Perhaps inevitably, Mark Zuckerberg's name came up, and his donation to their Kickstarter drive. Mike volunteered that he thought they should send a special thanks to Zuckerberg for his donation.

"We should be gentlemen about it," Mike said. That set off a mini-debate about the wisdom of having any contact with Zuckerberg. It came to a close when Komisar remarked that whatever the risk of getting in touch with him, there was none associated with simply ignoring him.

Max and Dan were furious at Mike's intervention. Ilya was sent as an emissary to tell him that he was not welcome to weigh in on business questions.

Their days were long, intense. Rafi, with ample family connections in San Francisco, had found an apartment, but Max, Dan, and Ilya had all moved into rooms at a cheap hotel that housed students and budget tourists. They hated it, but none of them did much beyond sleeping there. Rafi joined a gym that Mike belonged to, and was disciplined about getting workouts and a full night's sleep. At the end of the workday, Ilya would take a short run through downtown San Francisco, and he inveigled Max, who had never worked out, to join him for mental health purposes.

It wasn't simply that they were a small group taking on a huge project. No one, not the wealthiest corporation, most powerful government agency, or a small crew of hackers, can reliably predict how long it will take to make a computer program. New York City gave out a $63 million contract in 1998 for a new timekeeping system. By 2011, the payments had climbed to $628 million, and the project wasn't finished. As of 1995, the Internal Revenue Service had spent ten years and $2 billion trying to upgrade tax computer systems, without success, according to *Dreaming in Code,* an account of software development by Scott Rosenberg. And it's not just government inefficiency; businesses also struggle to execute grand software plans. Rosenberg described how McDonald's spent $170 million on a failed program to track food in thirty thousand stores. In some respects, such failures made the complications in the rollout of the Affordable Care Act in 2013 look minor.

In a classic work on the subject, *The Mythical Man-Month,* published in 1975, the author, Fred Brooks, explained that throwing people at a software project was counterproductive. When a job was running late, adding new people to it only delayed it. Each new programmer had to start from scratch, learning the logic already in place; teaching those concepts would consume the time of the incumbents.

A project that, in fanciful theory, should take one person three months would not be finished in one month if three people worked on it.

"Nine women can't make a baby in one month," wrote Brooks, who had managed software development for IBM.

He offered practical advice for programmers. "Plan to throw one away," Brooks wrote. "You will anyhow."

Standing outside the Pivotal offices on Market Street early one morning, Max saw a man with mad-professor hair, in sandals, approaching them. It was Bill Joy. They greeted him with awe, and then broke the news: they couldn't get upstairs. It was too early and the building wasn't open yet. They had been so worried that Joy, with his deep technical chops, would throw harder questions at them than the venture capitalists and news reporters that they had not considered the logistics of meeting so early.

Joy, who had made some of the most significant breakthroughs in modern computing while packed into a near closet at Berkeley, was not fazed by the setting. They stood chatting about the elements of Diaspora. He reported back to Komisar. "We met on the sidewalk. They were wonderful," Komisar recalled Joy saying. "They need to peel the onion, one layer at a time, and not just drive to the center."

That was the last they saw of Bill Joy.

Tuition for the design class was just about ten thousand dollars, and the check was signed "Casey Grippi."

Janice Fraser looked at it and tried to place the name. She knew that Diaspora had a Daniel Grippi, and Max and Ilya and Raphael.

"Who's Casey Grippi?" she asked.

"That's my dad," Dan said. "He's handling the money for us."

The guys had been sent to Fraser by Mitch Kapor and Rob Mee. The entire Bay Area was filled with people who were ready to explain the gospel of tech start-up, none more so than venture capitalists who had, through accomplishment or delusion or their proximity to seams of investment gold, come to believe that they knew what they were doing.

Fraser was not another fish in the start-up capital pond, taking no stake in the companies. She was a design guru, a kind of mentor for hire. For ten to twelve weeks, she ran seminars to help young entrepreneurs find ways to build a product that wouldn't blow over in a stiff breeze. Diaspora was one of five start-ups in her summer 2010 class. Designers could come and go, the pixels could be moved all over the screen, but as crucial as how a thing looked, that was the simplest part of the endeavor. Discerning its purpose was far more difficult.

"A lot of people think that design is the skin. Pretty pixels," Fraser told her classes. "That's not to diminish the importance of pretty pixels, but it's not enough. It's necessary but not sufficient."

At the first session, Fraser moved to each group to hear them explain what they were doing. She listened to Max tell the genesis story—the Moglen speech, the pizza-fueled coding sessions, the Kickstarter rampage. Then she popped her North Star question.

"Who," she asked, "are you building this for?"

Well, Max said, privacy on Facebook sucked. They were going to make it possible for people to have encrypted conversations. And the users would have the ability to take their data with them. There would be no question about who owned their personal information.

Dan had some ideas about taking away the limits on what a user could do—making it possible for a picture to be bigger, or to have a page of type.

Who, Fraser asked again and again, was Diaspora for?

"Is it for PLM?" she said.

That was Fraser's shorthand for People Like Me. Or people like Max, Dan, Ilya, and Rafi—distinctive personalities, unquestionably, but all members of the same food group, computer nerds. And proud of it. Were they building Diaspora for other guys who liked to hang out in the computer club and play online games? Google was an instance of something that was built for PLM. But that worked as a business because so many people felt they had to scratch the same itch of searching the Internet.

Gently, Fraser pushed her fingers through the holes of their work plan. They were working on creating a messaging function? Pictures would be shared? So? All the features wouldn't necessarily mean a thing unless they had an idea of who was going to use it. "Every application is one person," Fraser said. "You have to satisfy one person before you can make a market."

For many years, the developers of free software prided themselves on a stripped-down aesthetic. Linux was a fine operating system, and could do as much or more than Microsoft's Windows, or Apple's OS X, but it required users to type strings of commands that in slicker products could be activated by clicking a single button. Plenty of free software was like a car with a manual transmission instead of an automatic one: the stick shift gave drivers a better feel for the car and the road, but required a level of attention to the motor and its gears that simply was unneeded in cars with automatic transmission. Apple had achieved its success with a slick, simple user interface, shorthanded as UI, which mediated the user experience, known as UX. To break free software out of its dowdy appearance, a company called Canonical had developed Ubuntu, a handsome, easy-to-use operating system based on a version of Linux known as Debian; its good looks were viewed with suspicion in some corners of the free-software communities. Not by the Diaspora guys, though. They had grown up with the clean dazzle and elegance of Apple products, and expected that everything would look good, and work simply.

But by itself, a great look meant nothing, Fraser warned. Without substance, a clear notion of who and what the product was for, a cool interface was the graven idol, the golden calf.

The right answer, or at least the truth, was that they were not building a product. They were creating a function, something that would make it possible to use the distributed, federated nature of the Internet for people to swap and share without paying a toll in data. That function was not a product that could be sold; it would be like trying to monetize breathing. Yet the four pizza-munching dudes were taking on the coloration of Silicon Valley, where ideas mattered only if they were the kindling for start-up businesses. The questions posed to start-ups were also put to them—interesting, provocative inquiries that did nothing to dislodge a premise that was not true and had little hope of becoming true.

At the next session, a week later, they offered ideas about trying to create a network where strong ties would thrive, and not be overgrown by the weeds of acquaintanceship.

"We're not just the un-Facebook," Dan said. "We're not trying to out-Facebook Facebook. That's not interesting to us."

Max said: "What is really interesting to us is how we can build a product around meaningful ties. We don't want to end social networking. We want to fix it."

He brought along a deck of slides from a presentation by a Google researcher on the nature of online relationships. The thrust of it was that the existing social networks—that is, Facebook and the dying star of MySpace—did not encourage strong ties. Granting the obvious business interest of Google in undermining Facebook's spreading empire of content that was walled off from the relentless indexing machines of Google, the point was beyond dispute. There was no calibration. What you posted for the benefit of the softball team about the after-game beer party was the same thing that your mom saw.

Max spoke a mile a minute. Dan barely uttered a sound. Fraser suggested they draw the problem. Whether her students were garrulous or not, Fraser had often found that when people were at least temporarily emancipated from expressing themselves verbally, they could convey a thought visually. So Dan and Max drew an iceberg, a waterline where the peak protruded, and the underground mass. That tip was what people revealed online, shown to their weakest ties, the tiniest, usually least consequential bits of ephemera. Most of a person's identity was unseen, below the waterline. The hoariest of clichés, but it covered the ground that they needed. The fundamental shortcoming of the existing social networks, at least as the four saw it, was that everyone got to see the whole online iceberg. It was a dimension of the privacy problem that was a subset of the larger concern that had been mapped by Eben Moglen.

"Wouldn't it be good if you could really share with your strong ties?" Fraser asked.

Rafi was, in many ways, the oldest soul of the group, poker-faced under virtually all level of provocation. A diligent coder, he nevertheless did not permit the ravenous appetite of the project to consume all hours of his life. One day in mid-August, he called in sick. A stomach bug had laid him low. He was gone for the better part of a week, giving him an escape from the Diaspora grind.

While Rafi was away, Max brushed up on theories about the optimal, realistic size of an individual's social networks. On Facebook, people could have up to five thousand "friends," a boundary so elastically promiscuous as to dilute the concept of friendship beyond recognition. Maybe, Max mused, Diaspora should emphasize quality over wholesale faint acquaintanceship.

Two decades earlier, Robin Dunbar, an English anthropologist, had studied thirty-eight primate species to see if the size of the neocortex—the "thinking" area of the brain—related to the number of stable social relationships the primate was able to maintain. Indeed it did. Based on the volume of the typical human neocortex, Dunbar projected that the mean number of relationships for people was 147.8. Roughly defined, that was how many people an individual could reasonably ask a favor of.

His number was heavily critiqued. Checked against real life, the research stood up: communities of Amish, the Bushmen of South Africa, Hutterites, Native American tribes, and military companies all formed into groups of about 150. Bill Gore, the inventor of Gore-Tex fabric, walked into one of his early factories and realized that he no longer knew everyone. From then on, he built parking for only 150 people at each site, to ensure that the factories would not get beyond what he regarded as a manageable level.

More intimately, the primates studied by Dunbar belonged to smaller "grooming cliques," a social function that, he postulated, was replaced in humans by language. Whether they were picking bugs out of one another's hair, listening to tales of golf shots, or going on shopping trips, people got psychopharmacological benefits from the contacts, Dunbar and others argued.

It was the study of human networks that gave rise to the concept of "six degrees of separation." In 1967, a psychologist named Stanley Milgram sent letters to random people in Omaha, Nebraska, and Wichita, Kansas, and explained that the final destination was a stockbroker in Boston. The object was for the people in the midwestern United States to mail their letters to someone they knew on a first-name basis who, in turn, might know the stockbroker. As letters moved along the chain, each person who received one was given the same instructions. The median number of "hops" was five and a half or six. Milgram published an article on the study in the premier issue of *Psychology Today,* and later in academic journals. His methodology was widely criticized for introducing various kinds of experimental biases, and failing to account for the high percentage of letters that simply dropped out of the study. He died in 1984.

Yet Milgram was not far off. As Facebook encircled the world in digital ribbons, it provided data for the largest study ever conducted of social networks. It showed how many paths were needed to connect any two people, assuming they were not friends. Researchers working with Facebook reported in 2011 that 92 percent of all Facebook users—then about 721 million—could be connected in a path of five people, and 99.7 percent in a path of six. The average distance in the United States was 4.3 people.

So, yes, Milgram was right: there were no more than six degrees or so of separation among humans. But so was Dunbar: meaningful relationships didn't extend much beyond the second degree. The same study found that the median number of friends an individual member had was ninety-nine, and while there were people with thousands of friends, most had fewer than two hundred.

After Rafi came back, the four of them discussed it in a conference room.

"There's research that shows 150 is the number of relationships a person can plausibly maintain," Max said. "I think that limiting it to 150 would be something we would do."

The other guys thought there was some merit to this view, but Rafi came from sprawling families on both sides. Given a few minutes, he could probably add up more than one hundred cousins by going out to the second degree. About ten days earlier, a woman approached Rafi in the airport in Portland, Oregon, when they were on their way back to San Francisco from the federated social web conference.

"Are you a Sofaer?" she asked Rafi.

She, of course, was another cousin.

Limiting the number of friends on Diaspora to 150? He'd never get past family members.

"I think that's a terrible idea," Rafi said. "I would never use that product."

It was the kind of argument that flushed them out of the software engineering cave and into a conversation about social values.

"You can't be private with a thousand friends," Max said.

Rafi made a face. He understood that. But the hard cap on how many friends? He foresaw tiffs and hurt feelings over who was in, who was out. "This could bring back high school," Rafi said.

Max suggested that Rafi's resistance to the limit was due to his stomach virus.

"And you'll agree," Max said, "when you get whatever was in the water last week out of your system."

"I wasn't infected by whatever bullshit you heard last week," Rafi said.

Late one afternoon, a tall and massive man with a shaved head stepped from the elevator on the third floor of 731 Market Street, and into the buzzing floor of Pivotal Labs. There was no receptionist. Instead, Yosem E. Companys beheld dozens of long tables and people plugged into headphones and laptops. No one seemed to be in charge, but the space had the focused energy of an anthill. He wandered from one end of the floor to the other, a city block, looking for faces he knew only from photographs.

A few times he asked people who didn't seem to be fully absorbed by their computer screens if they knew where the guys from Diaspora were. No one did. Of course, the four had no idea what Yosem looked like. Of the thousands of strangers who had gotten in touch with them, his messages had been the most resonant. Not long after he said he was going to bring them to the attention of Randy Komisar, they heard directly from the VC. He introduced them to a half dozen other people. All of this was done by e-mail, while they were in New York and Yosem was on the other side of the country in Palo Alto. When they moved out west, the fellows had invited him to dinner in thanks for all the introductions and spadework he had done.

He arrived a bit early. So he sat down at an open computer and ran through his e-mail for a little while, then wandered down to one of the conference areas and spotted someone he recognized: Ellen Pao, a Kleiner Perkins partner who worked with Randy Komisar. She was just finishing a meeting with the four.

The guys all greeted him warmly and strolled a few blocks to a restaurant Ilya liked and settled into a booth.

A PhD student at Stanford who had degrees from Yale and Harvard, Yosem, thirty-five, was not only more than a decade the elder of any of the Diaspora group but he also had been marinated in entrepreneurial culture as an academic and a professional. He had worked in investment banking at Merrill Lynch and Goldman Sachs, researched intelligent work systems for General Motors, and brand management at Procter & Gamble. He was involved with an online group called Liberation Technology that shared news about ways for political activists to communicate while minimizing harassment. Believing in their cause, he was keen to help make Diaspora a reality.

"How did you even hear about us?" Ilya asked.

It was quite a story. Yosem and friends had been fed up with Facebook and its policies for quite a while, and believed that a distributed network, like the one the Diaspora crew had in mind, was the best answer. For much of the previous year, Yosem had been fighting a brain tumor, and stopped work on his doctoral dissertation for surgery in December 2009. Doctors tinkered with his hormones and prescribed heavy doses of steroids, adding bulk to his large frame. To occupy himself during the lengthy recovery, he had been having deep online discussions about how a peer-to-peer social network would function—precisely what Eben Moglen had been advocating. It was then that he discovered the plans for Diaspora, which he immediately passed along to Randy Komisar, who had taught a class on entrepreneurship at Stanford. Yosem had been his teaching assistant. He had also pulled at the sleeve of Tim Draper, a third-generation Silicon Valley venture capitalist who had backed ventures like Skype and Hotmail. Draper, too, asked to hear more about Diaspora. Yosem was startled by the instant interest both Draper and Komisar had shown.

Every week, Max told him, they were meeting with one of two people at Kleiner Perkins, either Randy or Ellen. There had been no discussions of funding as yet, but it was clear they were interested. "Time with VCs is more important than money," Yosem said. "There's an old saying about VCs in Silicon Valley: ask for money and get advice. Ask for advice, and you'll get money."

After a few minutes, Yosem, who taught graduate engineering students at Stanford, realized that the fundamentals of venture capital financing were total mysteries to these kids who had never been outside a college campus. He started with the basics: how the shares of a company were diluted through various rounds, with early investors being rewarded for taking the most risk by getting the biggest stake for the smallest investment.

Ilya started taking notes on a napkin.

Randy Komisar had been pushing them to build a private network for friends and family. Yosem thought the idea had merit. Social network theory, which in sociology predated digital networks, was one of his academic interests.

"The research shows that the average inner circle is about five people," Yosem said. "Why don't you create the social speed dial? Limit the number of people in your network to nine."

Rafi grimaced, but said nothing.

The idea, Yosem said, was that anyone could follow your postings, but private conversations would be seen by only nine people. The idea intrigued them. It was a significant way to distinguish themselves from Facebook.

The design of the network absorbed them, almost to the exclusion of any thought of how such a network could plausibly sustain itself. To Yosem, they seemed to have given no thought to a business plan.

Perhaps, he suggested, they might consider operating as a not-for-profit—to find some kind of structure where they could build out Diaspora without the pressure of having to make money. But they did have some vague notions of a way that Diaspora could become a business.

In their initial e-mails, and in conversations with Randy, they had all discussed the idea of Diaspora providing a paid service of hosting "pods," individual servers that were under the ultimate control of the users but a task that many ordinary web surfers would find beyond their technical means. By the time they met with Yosem, all they thought about was building the thing. It seemed to him that they had bought into a piece of Silicon Valley lore that divorced business realities from digital acumen: they appeared to believe that all they needed was a working product, or something to demonstrate, and they could leave the business plans to the

funders. Since they were taking advice from Kleiner Perkins, Yosem did
not argue the point.

Indeed, the Diaspora Four were delighted to have someone who just
seemed to want to help them in concrete ways. And Yosem was thrilled
to find people eager to build the kind of social network that many people
wanted. "I was looking for someone to pick up the mantle so that I could
become their evangelist," he said.

Casually, he offered to prepare a case study of Facebook, as well as a
list of potential competitors, a remark that meant nothing in particular to
the guys until a few days later, when a fourteen-thousand-word e-mail
arrived from him. It was an account of the rise of Facebook, tracing its
roots to initial efforts, among others, at Williams College in 1989, when
Mark Zuckerberg was not yet five years old, and a decade later by a former
Tiananmen Square protestor who had moved to Cambridge. The first big
commercial social network site in the United States, Friendster, preceded
MySpace by a year and Facebook by two years, but was hobbled by its
popularity; the demand outpaced the ability of the technology available
in 2002 to keep up with the capacity of the servers. (The company was
underwritten, in part, by Kleiner Perkins.) By the time Mark Zuckerberg
offered Facebook to a broad audience in 2004, the technical infrastructure had improved. Not only was Zuckerberg talented, farsighted, and
fast moving, but his timing turned out to be perfect.

As the dinner wound down, Ilya took out his Maker's Notebook,
which he carried everywhere to track ideas. He slipped the napkin of
notes into it.

"Someday, when we are rich and famous," Ilya said, "we will pull out
this napkin and say, 'This is how it all started.'"

In early August, the team realized Diaspora 1.0 was hopelessly nerdish.

They demolished it. As Brooks had prophesied in *The Mythical Man-
Month,* they had to be ready to throw away one version, or more.

As they rebuilt, they were quicker. Roadblocks became manageable.
Getting photos into the stream had seemed, initially, to be beyond their
competency.

"Next year, maybe we will have photos," Dan said. "Then one week,
we were, like, whatever. And we had them."

Ilya cackled. "We thought that was never going to happen," he said. "Maybe everything seems hard until you try it."

That was another lesson from the programming gurus: find a problem that interests you, and you will solve interesting problems. The Diaspora team was wrestling with a challenge that they cared about, in an area of life that was fundamental to their generation. And as daunting as the notion of designing a network had been, they had found that many building blocks were widely available. Eben Moglen had been right: Facebook was a bunch of doodads written with the PHP code that made web pages come to life, with things like pictures and status updates. By their more cherished measuring stick, innovation, Facebook was pretty minor.

"Facebook is not that genius," Dan said. "It's like a stupid website. It's a lot different than the stuff Steve Jobs is doing."

The basics seemed simple, Ilya said: "Social networking—I can post photos, I can post messages. And oh, I can post location. In itself, it's just two fields in a database somewhere."

He paused for a moment, reflecting on the hubris, or possible ignorance, of his remark. "Maybe it's that nothing seems hard when it's already there," Ilya said.

A few people at Pivotal who had taken them on as mini guidance projects spent two days going over their first version of the site, and pushed them to identify three classes of users.

They came up with, at one extreme, people they called "beards"— free-software users—a class of übergeek, and, at the other end, "girls." The girls and the beards were polar opposites in this rendering of humanity. (The third class, lodged in the middle, were people who used free browsers like Firefox rather than Internet Explorer, and were glad to employ tools that they did not have to devise themselves.)

"We have to have girls if we are going to succeed," Max said.

Two of their cherished details, though, were arduous for all but the most devoted geeks: encryption of data and running an individual server. The original concept of Diaspora was that each user would have his or her own server, of the kind envisioned by Moglen for the freedom box. These individual servers would be called "seeds" (and thus the aptness of the name Diaspora, with its Greek root evoking "scattering" and "spores").

While the hardware for the seed servers existed, the day when it would be a simple plug-in and setup was some time off, awaiting the development of a stack of software that would include a social network like Diaspora.

Moreover, Max did not want Diaspora to be just another piece of software to end up in the freedom box. "We don't want to be part of Moglen's army," he said. Max had been put out by Moglen's statement in the *Wired* article that he was not going to pick a favorite among projects working to create an open-source alternative to Facebook. "We thought it was heresy," he said.

To get wide acceptance of Diaspora in the short term, the group recognized that servers would probably have to be hosted by schools or companies. The notion of seed servers was supplanted by the idea of pods.

Then there was the question of making encryption manageable. Both Rafi and Ilya had been deeply invested in it as an intellectual and political challenge.

"We're in charge of the tinfoil hats," Rafi said.

Encryption at one level or another had been practiced by armies and lovers for centuries. The point was to keep confidential communications from being hijacked by encasing messages in codes that, in theory, are known to only the sender and the recipient. Until the early 1990s, most strong encryption was under the control of government agencies. An antinuclear activist named Phil Zimmerman devised what he half-jokingly called Pretty Good Privacy in 1991, so that politically engaged people could communicate on Internet bulletin boards without being watched. The code quickly made its way around the world. Two years later, Zimmerman became the subject of a criminal investigation into the exporting of munitions without a license: encryption that used keys larger than 40 bits was classified as a munition under U.S. regulations, and PGP keys were at least 128 bits in size.

With help from an aroused community of academics—how could algorithms be controlled?—and lawyers, including Moglen, Zimmerman came up with a plan to circumvent the ban. The law might see a 128-bit encryption key as a munition. As pages of 1s and 0s on paper, though, the code was a physical thing called a book. The entire PGP code would be published by MIT Press. Anyone could buy it, tear off the covers, scan it into a computer, and encrypt away. There were no export controls on books, or tearing out the pages, or scanning them. The investigation of Zimmerman was dropped, and PGP continued its global migration. In the United States, federal law enforcement officials took the position that for public safety, encryption should be regulated by the government. To create impermeable hiding places was, in the view of the FBI director, Louis Freeh, a development that would help terrorists, criminals, and others seeking to damage society. Starting in 1993, the Clinton administration encouraged the makers of secure communication devices, like certain phones and fax machines, to voluntarily incorporate the "Clipper chip," an encryption tool developed by the National Security Agency. The decryption key for the Clipper would be held in a kind of legal escrow by the government, and was to be used only when it had permission to unlock encrypted conversations on the devices. The industry resisted. Without voluntary compliance, the Clipper program failed.

Nevertheless, generations of government authorities continued, with little public knowledge of what was going on, to intercept communications on the Internet. In 2002, John Poindexter, a presidential adviser, proposed a "Total Information Awareness" plan to scan phone calls, e-mails, and travel and financial records from across the world. That seemed too drastic a step even in the months after the attacks of September 11, and it was abandoned in 2003. Or so it seemed. New programs, called Prism and Bullrun, did essentially the same thing, with much greater effectiveness and despite varying levels of cooperation from the Internet companies. In 2013, Edward Snowden revealed that the National Security Agency had gotten access to traffic on Google and Yahoo! without the knowledge of either company. Even though those companies have complex security walls at their data centers, both virtual and physical, the information is not encrypted when it moves across the cables that are the backbone of the Internet from one center to another. "People knowledgeable about Google and Yahoo's infrastructure say they believe that government spies bypassed the big Internet companies and hit them at a weak spot—the fiber-optic cables that connect data centers around the world and are owned by companies like Verizon Communications, the BT Group, the Vodafone Group and Level 3 Communications," Nicole Perlroth and John Markoff reported in the *New York Times*. "In particular, fingers have been pointed at Level 3, the world's largest so-called Internet backbone provider, whose cables are used by Google and Yahoo."

As the digital ecosystem grew—"modern cars are computers we put our bodies in and Boeing 747s are flying Solaris boxes, whereas hearing aids and pacemakers are computers we put in our body," the writer Cory Doctorow observed—vast resources were expended by businesses, criminals, governments, scientists, and hackers to seize control of at least part of the space. Measure was met by countermeasure. A computer user might lock a machine with a password, but could still face catastrophe by opening an innocuous-looking piece of e-mail and clicking on an attachment. The poisonous software secreted within those attachments would install a keylogger—a spy tool that tracks, records, and transmits every stroke entered on the keyboard, thus stealing passwords, account names, and so on.

For certain hackers, encryption was a fundamental tool. But it was also a pain in the neck. The easiest part was the creation of lengthy encryption keys, which could be automatically generated. From there, it got complicated. Both sides on an encrypted message had to have the same key in order to unlock it, and the key had to be shared by trusted correspondents, generally in face-to-face encounters.

The guys couldn't find a way to make it feasible for Diaspora, and settled for SSL, a form of encryption used by banks for transactions with customers. "It makes me ideologically sad," Ilya said. But he knew that to succeed, what they would present to the world in September had to be something that ordinary people would be able to use without breaking a sweat. Their project could not be tailored to software obsessives.

"We have to have girls if we are going to succeed," Max said.

"Between now and October," Ilya said, "we need to get actual girls and say, 'What would get you to use this?'"

"Chocolate," Dan mused.

It was well and good to be programming at all hours of the days, but what about the T-shirts they had promised their Kickstarter supporters? They owed thousands of them, in specified sizes, to people who had contributed; other donors were owed stickers or various doodads. The DVDs with the code would have to wait until they'd finished a full draft. The most vexing problem was the microswag.

Max had an aunt in the T-shirt business; and his high school buddy Dan Goldenberg, who had spotted the girl on the bus after the NYU graduation reading the Diaspora Kickstarter pitch, was moving out to San Francisco and didn't have a job. For three months, Dan found himself serving as the head of the Diaspora fulfillment department—getting all the stickers and stuff out the door.

# CHAPTER NINE

Twice every hour, Caltrains departed from the station at King and Fourth Street in San Francisco, heading south down the peninsula. It is a trip known to many young entrepreneurs: a ride of less than an hour that would bring them to the gleaming offices of venture capital in Silicon Valley. On the twentieth of August, three of the Diaspora Four boarded the train and headed south; Dan had gone back to New York for a family wedding.

Unlike thousands who had traveled this path, they were not going to one of the celebrity success stories of the valley, or to the vaults of venture gold.

They were headed for a building in Mountain View that, by the summer of 2010, housed one of the most potent forces in modern cyberspace: the headquarters of the Mozilla Foundation. No organization had been more successful at bringing to life the values that inspired the Diaspora project. When it came to long-shot, popular, valorous digital causes, Mozilla just about wrote the book. It began, roughly, six years earlier.

A winter day in 2004 in Mountain View, California. Winifred Mitchell Baker walked around to the back of an office park, to an ordinary set of glass doors.

Two men were standing at the entrance.

"Hi," Baker said. "Can I help you?"

"We're here to see Mozilla," one of the men said.

Mozilla was where Baker worked. But it was far too early in the day for appointments in this obscure corner of Silicon Valley.

"Seeing anyone special?" she asked.

"No," the man said. "Just Mozilla."

"Well, why are you here? What are you doing?" Baker asked.

"I'm a Firefox supporter, and I came here to see the Mozilla offices," he said, stepping toward the door, "and there's my name—there."

Taped to the door was a two-page spread from the *New York Times*. It was an advertisement. One of the pages introduced Firefox 1.0 as "the free, open source web browser from the Mozilla Foundation that lets you surf faster and more efficiently and helps avoid annoying pop-ups and spyware."

The other page, where the man was pointing, was a gray blur; on close inspection, one could see names, in tiny print, running down its full length.

At the bottom was an explanation for the cloud of names.

"This message has been brought to you by the thousands who contributed to the Mozilla Foundation, a non-profit organization dedicated to promoting choice and innovation on the Internet."

Baker smiled.

"Let me open the door," she said, "and then you can come on in and wander around. Until you feel full."

Three years earlier, Baker—known universally as Mitchell—had stood up from the wreckage of a company called Netscape, dusted herself off, and come back to finish what she and others had started. In its earliest days, Firefox was just a postscript to a saga that embodied the genius and folly of the early Internet age like no other.

In time, though, it would emerge as the single most powerful defender of the World Wide Web as a public commons, a bulwark whose growth was essentially uncharted in the business news pages but that was a defining force in the digital ecosystem.

The network of networks called the Internet had existed since the 1960s; the web, an entirely different creature, was brought to life in 1989 and 1990 by Tim Berners-Lee and Robert Cailliau, computer scientists working at the CERN particle physics laboratory in Switzerland. The lab's researchers and data were spread around Europe and the rest of the world; vital information might be stashed in a computer and needed by people a floor, or a continent, away, and they would have no way to put their hands or eyes on it.

In an "act of desperation," Berners-Lee would later say, he and Cailliau assembled existing pieces of technology, like protocols that let computers speak to one another, a technique of indexing called hyperlinking, and an address system known as URL—uniform resource locators—based around domains, like postal codes in the physical world. He called this collection of technology the World Wide Web.

When it was launched in August 1991, Berners-Lee said: "The WWW project was started to allow high energy physicists to share data, news, and documentation. We are very interested in spreading the web to other areas, and having gateway servers for other data. Collaborators welcome."

Collaborators arrived at a brisk pace, as Berners-Lee and Cailliau sought no patent for their WWW thing. Many wanted to expand the web beyond the bare bones arranged by Berners-Lee, a system of static, unchanging documents that one person could summon from a website. Brilliant as the first web was, the tools for sharing and looking at information were obtuse, nonintuitive. The web was little known.

Then came Mosaic. Working in a supercomputing center at the University of Illinois in 1993, Marc Andreessen and Eric Bina developed the code that made it easier to use the web to explore the world outside high-level research facilities. Seen through Mosaic's shimmering window, the web could be alive with pictures and graphics, a dynamic ecology of give-and-take, call-and-response, like human dialogue. Compared with the primitive techniques then in place that required typing hard-to-remember commands, Mosaic allowed the users to gaze on a panoramic wall of glass, instead of squinting through a peephole. By October 1994, Mosaic had increased the traffic on the Internet dedicated to hypermedia browsing by ten-thousand-fold, Gary Wolf reported in *Wired*. Anyone connected to the Internet through a modem could use the browser. No longer did people need to subscribe for a fee to curated services like America Online, CompuServe, or Prodigy to surf the web, nor were they limited to e-mailing just between other customers of the same service. Mosaic could be installed on any computer, instantly demolishing the walled gardens operated profitably by AOL and its competitors.

"Mosaic is not the most direct way to find online information," Wolfe wrote. "Nor is it the most powerful. It is merely the most *pleasurable* way, and in the 18 months since it was released, Mosaic has incited a rush of excitement and commercial energy unprecedented in the history of the Net."

The possibilities were vast: "Long-frustrated dreams of computer liberation—of a universal library, of instantaneous self-publishing, of electronic documents smart enough to answer a reader's questions—are taking advantage of Mosaic to batter once more at the gates of popular consciousness. This time, it looks like they might break through. Mosaic is clumsy but extraordinarily fun. With Mosaic, the online world appears to be a vast, interconnected universe of information. You can enter at any point and begin to wander; no Internet addresses or keyboard commands are necessary. The complex methods of extracting information from the Net are hidden from sight. Almost every person who uses it feels the impulse to add some content of his or her own."

The word "browser" conveys its early function of thumbing through information, a tool for glancing. That term persisted, even as the tool known as the browser became the skin of the cyber age, a membrane that gave every person access to the Internet, and the Internet access to the person. The workings of the browser mirror existence: From the simplest amoeba to the great primates, life sustains itself behind cell walls, membranes, that use alignments of molecules, their + and – signs, to regulate metabolic transactions. They control what comes and goes, what the outside world gets and what it gives. In the online world, the browser, with its arrays of 1s and 0s, determines the points of equilibrium between an individual and the web, what data pours out when information is taken in. How that arithmetical membrane works, the conditions under which it is permeable and where the balance of power resides, is a matter of far more than technical, academic interest.

With Mosaic, the calendar of change in the human species, once measured in epochs, had been compressed to the pellet of a few months. The most social animal had evolved an entirely new way of playing and learning, fighting and flirting. Surely, there were businesses here. The creators of Mosaic, Andreessen and Bina, partnered with the entrepreneur Jim Clark in 1994 to develop a commercial browser that they wound up calling Netscape Navigator, dodging legal issues with the University of Illinois over the Mosaic name.

On August 9, 1995, Netscape offered shares in the company for public sale, initially priced at $28. At the end of the day, they were selling for $58, making the company worth $2.9 billion in the market. By the end of the year, with the shares selling for $171, the value had more than tripled. But what was the business? Where was the money? Netscape sold the browsers to businesses and gave them away to individuals; when the browser opened up, it brought the user first to a Netscape home page, and it was thought that might be prime advertising real estate. In truth, the company had virtually no profits and no plan for getting any, but there was mesmeric power in sky-high piles of capital.

"In 24 months, the Web has gone from being unknown to absolutely ubiquitous," Mark Pesce of *ZDNet* wrote in "A Brief History of Cyberspace," published in October 1995. With vast amounts of money, Netscape hired every smart person who walked through the door. All paths to the richness of the web ran through the browser. If people with money were treading those paths, there might be ways to extract some.

The rise of Netscape unleashed huge amounts of venture capital in pursuit of the next big thing, whatever that might be. Something with the web. Anything with the web. Hope-addled, lottery-playing, clueless investors surrounded this web thing, the core of a genuinely new world, with an atmospheric bubble of hallucinatory gases.

In Redmond, Washington, home of Microsoft, the last big thing, Bill Gates followed these developments. Until then, Microsoft had focused on writing the software that ran personal computers, and controlled how data was crunched and words were processed. All of it was done in glass and aluminum boxes that sat on desks. As those boxes could become portals to the Internet, the horizon was infinite. To get there, Microsoft created a browser called Internet Explorer, a descendant of the original Mosaic, under a license from the University of Illinois.

Thus began the browser wars. Microsoft wound Internet Explorer into the gears of its Windows operating system; Windows itself was bundled into personal computers that were being bought by the tens of millions. So Internet Explorer was "free," as long as you bought the two-hundred-dollar operating system inside the twelve-hundred-dollar computer.

Browser usage is an arcane subject, with many imperfect metrics, but all of them showed that after Microsoft entered the game with Internet Explorer, usage of Netscape dropped from more than 90 percent of web surfers in 1995 to around 60 percent in 1998. It was losing ground every minute.

As with any membrane, the browser's protective ability was an essential quality but its ability to fend off malignant hackers begins to erode on the day of its release. To keep up with security threats, not to mention adding new features, meant that browsers had to be in a constant state of rebuilding. Netscape fell behind. The steady rise of revenues stopped. It could not sell a Netscape browser if Internet Explorer was "free." In time, Microsoft's tactics would be the subject of an antitrust lawsuit from the U.S. Department of Justice and regulatory responses from European authorities, but those cases would take years to resolve.

Meanwhile, Internet Explorer rose like a rocket. In response, Netscape, having rushed out flawed new versions of its browser, made a drastic decision in 1998. No longer would it develop new versions of Netscape in secret. The company turned over its browser code to an internal organization that would build new versions in an open process similar to what had been done with the creation of the Linux operating system, drawing on the contributed wisdom of programmers around the world. That internal organization at Netscape was called Mozilla. (The code name was created at a marketing meeting when someone mentioned crushing Mosaic, an Oedipal, or at least patricidal, act, considering it had been the direct ancestor of Netscape. But by then it formed the core of Microsoft's Internet Explorer. A developer named Jamie Zawinski fused the name of a Japanese movie monster, Godzilla, with the word "Mosaic." The name Mozilla stuck.)

Later, in November 1998, Netscape was bought by AOL, one of the curated gardens whose walls had been rendered moot by the emergence of the browser. Even in Netscape's diminished state, the initial deal, based on stock values, was $4.2 billion. It made no sense. Four months later, when the sale closed, the stock prices made it worth $10 billion. Helium was keeping Netscape aloft. Its prime assets were in the hands of what was then known as the Mozilla Organization, the skunkworks project that was developing the browser. The 1s and 0s of the code by then were openly available on the Internet for anyone to see and fiddle with.

Mitchell Baker had been with Netscape nearly from the outset, and was sent to oversee the Mozilla project. She wrote the license agreement that spelled out the terms of the barter between those who contributed elements of code and the company that was assembling the intellectual property of multiple intellects. Declining to name herself chief executive officer, she chose the title "Chief Lizard Wrangler."

The contributions from the open-source community were brilliant, banal, and all points in between. The disputes over the finer points veered toward the deranged, flaming wars on chat sessions. Jamie Zawinski, the same developer who came up with the name Mozilla, characterized the process as "mass nonconsensual psychiatric care." The level of vitriol online—not just the list-serve discussions for Mozilla—was so profound that Mike Godwin, a lawyer with the Electronic Frontier Foundation, wrote Godwin's Law of Nazi Analogies. It held that "as an online discussion grows long, the probability of a comparison involving Nazis or Hitler approaches One."

In 2001, AOL merged with Time Warner, the publishing and entertainment conglomerate; by August, the new company had started layoffs of Netscape employees working on the Mozilla Organization project. Microsoft had won. Mozilla produced a single, crummy browser used by virtually no one anymore. But within Mozilla, a dissident project, distinct from the original skunkworks, quietly started to take shape. It, too, was a browser, but different from what the world had already seen. The internal name was Firefox.

Two years later, by the summer of 2003, AOL decided that it was going to jettison the remnants of Mozilla. Yet Mitchell Baker, laid off in August 2001, was still there. Even though she had not been paid by the company in two years, she had been running the dissident project. How had that happened?

On the day she was fired, it turned out, Baker had been on her way to a meeting of a foundation that was promoting the use of open-source programming and was run by Mitch Kapor, the creator of Lotus. Kapor had long thought that a Microsoft monopoly on access to the Internet was a poor state of affairs. But he knew, too, from his own work that Microsoft's dominance had created an investment vacuum, with venture capitalists tremulous about funding inventive people trying to gain ground in any of the areas controlled by a monolith. Open-source development, Kapor reckoned, might be a way around that vacuum. The intriguing lesson of Linux had shown it was possible. So he started the Open Source Applications Foundation to develop a few pet projects.

Mitchell Baker, who had never met him, arrived at the foundation meeting in 2001 fresh from being fired.

"What are you going to do next?" Kapor asked.

"I'm going to keep going back," Baker said.

Kapor put her on the payroll of his foundation to subsidize her. Baker simply kept showing up at the Mozilla Organization as the volunteer Chief Lizard Wrangler. Somehow, this state of affairs continued for two years. In 2003, Baker told Kapor that Firefox was actually making good progress, but it was clear AOL was not going to sustain it. There was no reason for a Mozilla–Netscape–AOL–Time Warner.

Kapor, an éminence grise of Silicon Valley, suggested to senior people at AOL that the company should avoid the embarrassment of being seen putting the splinters of Netscape out with the trash. It could do this by setting up a Mozilla foundation to take over the development of the browser. This idea took hold at AOL, which agreed to do that and stake the new foundation to about $1 million, scarcely a rounding error on the books of the new corporate behemoth. Kapor goosed them, and it went to $2 million. Kapor also donated several hundred thousand dollars of his own money.

The Mozilla Foundation was created. Jettisoned from its corporate Frankensteins, lost to the sight of the venture capital markets, the Mozilla project called Firefox continued its ridiculous journey. In the saga of how the browser has evolved in the service of humanity, the free fall of Netscape, spectacular as it was, is not the most important part. The escape capsule bearing Firefox was far more significant.

At the time the Mozilla Foundation was created in 2003, Mitchell Baker was forty-four, and a strikingly egoless personality. A lawyer and a linguist, she possessed an emotional economy that kept her from burning up resources in needless conflict. After seeing a trapeze performance in a park where she'd brought her son, she decided to give it a try. She spent weeks smothering her fears, then began lessons. At an age when most people have gotten very attached to standing on firm ground, she was flying three times a week.

Mozilla existed in a crevice of Silicon Valley that she found comfortable, outside the demands of the normal investment markets, but able to move quickly and nimbly, like the best start-ups. It was obvious that there was a need for what they were doing. The world's dominant browser, Internet Explorer 6, was plagued with infernal pop-up ads. Worse, a tool within its code could turn Explorer into a vector for viruses. All in all, it was widely and deservedly hated. With a de facto monopoly, Microsoft had little incentive from the market to improve it, and, in fact, had essentially disbanded its development and sent the programmers off to different projects. Because the Explorer code was closed, people from outside the company could not look at it and identify the problems. They could only experience them. And yet, as Baker told *Wired* magazine: "There was no interest in the venture capital world in funding another browser. Netscape had died trying to fight Microsoft. Who would ever try and compete in that space, especially after the browser had been done away with as a separate product and combined with the operating system? So in that setting, many of us were eager to interact with the web but the only available tool for doing so was low-quality, poor-performing, and a security risk."

Mozilla, the foundation and its gestational-stage browser, needed a new home. Baker sublet office space in Mountain View from a tech company run by a Netscape alumnus whose business was contracting after the tech bubble sputtered. As necessary as another browser was, though, no one was paying much mind to the Netscape/Mozilla refugees who fought on, seeming like Japanese soldiers on remote Pacific islands after the end of World War II, unaware that their cause was long lost or unwilling to accept that reality.

"The next big thing," said John Lilly, who became a Mozilla executive much later, "is always beneath contempt."

The invulnerability of the incumbent big thing was not, it turned out, a permanent condition. In September 2003, the federal government created the Computer Emergency Readiness Team, a task force on cyber-security issues. In a series of communiqués over the next year, the Emergency Readiness group declared that Internet Explorer was dangerous because a portion of the code it used, called active X scripts, was especially vulnerable to exploitation. A hacker could secretly take over a computer, infect it, send millions of spam e-mails, break into bank accounts.

This was bewildering news: Internet Explorer had 95 percent or more of the browser market. What could the average person sitting at a computer do? Steer clear of Internet Explorer, the feds said.

"It is possible to reduce exposure to these vulnerabilities by using a different web browser, especially when browsing untrusted sites," the advisory said. But even if a different browser was used, the danger continued as long as Explorer was part of the operating system.

At the same time that these dire findings were coming to light, the early versions of the Mozilla browser, Firefox, were getting raves from test-drivers. It was lean and fast, offered tabs in the browser—then unknown, but which have since become a universal feature—and could stomp out the pop-up ads that were both irritating and potentially disruptive to the computer. The browser was built so that developers could contribute "add-ons"—handy features for any kind of task, from managing pictures to sending out an alarm. The Mozilla staff of fourteen—the lead engineer, Ben Goodger, was twenty-four—oversaw contributions from volunteer developers around the world who were making a browser that served the interests of the users.

As the Firefox digital barn raising neared its end, there were calls to celebrate the moment publicly. Mitchell Baker was skeptical, perhaps from the same emotional tropism for reticence that, over the years, kept one of the most powerful forces in technology also one of its least known. But the quirk of a single person did not stop the community of quirky people who had created Firefox. They would put out an ad. Fine, Baker said, just don't expect us to pay for it.

Hardly anyone involved had experience putting together a newspaper ad. Chris Messina, a recent graduate of Carnegie Mellon working as a volunteer designer for Mozilla, agreed to take on the job. Word spread among the legion of volunteers: anyone who pitched in would be named in the ad. They needed only $20,000. Within ten days, they had pulled in $200,000, much of it in $5 and $10 contributions.

Like Diaspora's startling $200,000 bounty in 2010, the Mozilla cup also ran over, almost to the identical dollar, in 2004.

Instead of a single page in the *Wall Street Journal,* they would take out a two-page spread in the *New York Times.* One page would feature the Firefox logo and a message about the new browser. The other would display a list of every one of the thousands of donors. Messina crashed four or five computers trying to render the ad for print. It took consultations with Adobe, the maker of the page-making software, to figure out how to get that much type into such a small space. The ad ran on December 16, 2004. Firefox had arrived.

"It was a lightning bolt to the world. We could build something usable," Messina said. "This was the canonical fuck-you to Microsoft."

Launch parties were being held on six continents. Then an e-mail arrived at headquarters from a man name Ethan Dicks, whose signature automatically included the weather in his area. It was -66°F, with a windchill of -101.9°F, at his station in the South Pole. So a launch party was held there, too: Mozilla Antarctica made the seventh continent, completing the global set.

Tech writers praised the speed of Firefox, its tabbed pages, the vanquishing of the pop-up plague, the more secure environment. It also had the whiff of rebellion. In the first month of its release, Firefox was downloaded 10 million times. Eben Moglen was fond of saying that when it came to free software, only one copy was needed. "We scale," he said.

It was a daring moment. "Open source is relatively well-known and understood today," Baker said in 2010. "That wasn't the case when we started. This is a pretty radical way of doing things. It means there is nothing secret about our product. You can take and use and reuse and copy the idea, any piece of what we do." (In time, Google began to work that way, a move that delighted Baker.)

People made Firefox better because they wanted to. A Mongolian expatriate living in Hamburg, Germany, translated the code into Khalkha Mongolian so that he could use Firefox to keep in touch with people at home. Volunteers translated it into sixty other languages, a global girdle stitched only with human capital.

A feature devised by a man in the Middle East also quickly became popular: a privacy setting. When it was turned on, no record was left on the computer of the websites visited. Makers of other browsers, including Internet Explore, Safari, and Chrome, all eventually developed versions of it. The privacy setting became known throughout Silicon Valley as "porn mode," even though the browser makers had suggested that users of the setting could have entirely wholesome purposes for furtivieness, such as shopping for surprise presents for family members. But a study by researchers at Stanford and Carnegie Mellon universities found that far more time was spent by private browsers on what it called "adult" sites than on, say, gift-shopping sites.

Snickering aside, the mode had not been devised as a way of covering the tracks of porn gawkers. John Lilly said the Firefox volunteer who had created the privacy setting lived in Teheran and was concerned that if his computer was seized by the authorities, they would have a record not of any interest in smut sites but in political ones, which would be far more transgressive.

Even a free platform, supported by the sweat equity of thousands of people, costs money to build, improve, refine. The going-away money that AOL had given Mozilla would not last long, and Lilly, brought into the project by Mitch Kapor, made financial sustainability his mission. The solution emerged from a box at the top of the browser screen. It turned out that the search companies were willing to pay to be listed there. To capture that money, the foundation set up the Mozilla Corporation, a for-profit subsidiary that pays taxes and uses its proceeds to support the foundation's work. Before long, a burgeoning company called Google was paying Mozilla Corp. about $100 million a year to make a Google query box the default for searching from Firefox browser pages; other search engines were listed below it in the same drop-down box.

In one year, a not-for-profit foundation whose sole asset was a browser that had been created by volunteers would realize more revenue than

Netscape, its for-profit ancestor once valued at $10 billion, had in its entire existence. By 2012, Firefox's annual revenue had climbed to nearly $300 million.

The net proceeds went into the foundation to continue developing new releases of Firefox, and advancing its mission "to promote openness, innovation and opportunity on the Internet." The Mozilla manifesto declared the Internet to be a "global public resource," and said the foundation was committed to protecting the public benefits.

When Firefox was launched in 2004, Rafi, the youngest of the Diaspora guys, had not yet started high school. The web had changed every day since then. What had been, in essence, millions of connected libraries was transformed into millions of theaters where every man, woman, and child could perform, a collaborative space of infinite flexibility for sharing video, music, data streams, multimedia graphics—and also to run applications. But during the early years of the twenty-first century, the absence of standards left that space completely up for grabs: companies created proprietary systems where, for instance, they could share video, but no one else could. Baker and Mozilla were potent forces in driving for a new open-source web language, HTML5, along with standards that baked in the interactive features. Before the decade was out, the new language became the foundation of the web. Faster than anyone could calculate, Baker realized, it expanded the rewards and the dangers of being connected to the web.

"The hidden cost of admission is this data about ourselves that we give up, and it's painless at the moment you give it up," Baker said.

"Some people say, 'I don't care, it's never going to be painful for me— I don't care if everybody knows all about me or if they're monitoring, I like the highly personalized ads. You know where I am at five o'clock on a Friday afternoon, because you're a website or you're tracking me? That's okay with me.' Then there are some people who are truly appalled by that."

Firefox offered users ad-blocking mechanisms and a tool called Lightbeam, allowing them to display in mesmerizing detail the tracking bugs active on a computer. "We're trying to build a certain kind of Internet where each human being is sovereign regarding their technology," Baker said.

Inevitably, the Diaspora team would end up at the Mozilla offices that summer. Aza Raskin, the twenty-six-year-old creative lead at Firefox, was eager for their visit. "We are the open-source organization most aligned with them," he said.

At the Mozilla podium, Max clicked through slides to illustrate his talk. "People all over the world were talking about a project that we were doing in a closet at NYU. We actually hit our goal in ten days."

Seated in the front row, Rafi winced. He knew what was coming next. Max continued: "Then Wednesday, May 12, the day I graduated from college. I was sitting in Yankee Stadium in the rain. I started to get texts."

Rafi didn't want to hear the creation saga yet again. It was something that happened to them, not anything they had done. Plus it was already August, and they were due to release their code in September. They were working all hours to hit that. Going back over the events of the spring, dazzling as they were, would not help them get the thing done.

Still, it was important to lay out how modest their plans had been, especially now that they were taking a different turn. They had initially thought, Max said, "we're going to make some software that other people are going to hack on."

Instead of two hundred contributions from friends and family members, the donations had come from more than six thousand people, most of them not members of the tribe of techno nerds. "A lot of normal people," Max said. "We're getting e-mails, 'When can I sign up?' 'When can I use Diaspora?'"

They were rescaling their ambitions. Diaspora would be available beyond the cloistered world of free-software developers. "We want to make something for everybody, for the average user," Max said. "Something that Firefox has done an amazing job at, really reaching an average user, someone who is not technical, and making them understand why it is the superior browser. People need that. Not just hackers."

This was a drastic shift in their ambition, but one that would be little appreciated by the public, which simply expected that they would be able to sign on and sign up.

Their goal was to build micronetworks that mirrored those in real life. "Public networks facilitate weak ties; we want to focus on building stronger ties," Max said. "Privacy is not about not sharing. It's about control. It's to enable sharing."

Rafi took the podium to discuss the internal plumbing of Diaspora, and how it was going to work to be federated. Social networks, for the most part, did not communicate with one another—there was limited ability to send messages from someone in one network to someone in another. "We will make it like e-mail," Rafi said. "It doesn't matter what e-mail server you're on, you're able to communicate."

The transparency of their process, he said, would demonstrate the value of openly developed software. It would help "normal people understand that they won't get taken advantage of."

Questions came from the audience. What about encryption?

Max quickly spoke on that. "Our goal is not to build systems that are anonymous," he said. "The group of people who want anonymity is smaller."

A pathway could be created for those who wanted that, he said.

Ilya, who had been sitting quietly until that point, entered the conversation to put down a marker for cryptography. At the moment, he said, it was hard to make it comprehensible and simple on the screen, in the user interface, the UI.

"We don't know how to work that into the UI right now," he said. "It's important to have encryption available."

Another question: "Are you still heads down, or do you want people to contribute?"

"Right now, we're heads down," Max replied, but in September they would open up the software. "We want lots of contributors."

They needed advice on getting ready for that daunting moment. "Any best practices?" Max asked. "We don't want to turn people away. We want to service them as much as possible. We are four dudes. We might get totally railroaded. We're going to be definitely very strapped for time. I'm kind of worried about what happens when we turn it on."

A Firefox staffer spoke: "I think you should be soliciting the community for more than just coders; also for people who get really familiar with the project, and someone who can triage for you. I think that's really what you're looking for first and foremost. You don't want to get overwhelmed."

"'Cause that's going to happen," Max said.

"You need a gatekeeper to prioritize," the Firefoxer said.

That wave of people wanting to help, and the prospect of being swamped by it, was on Ilya's mind.

"That's one thing that definitely happened to us after the *Times* article," Ilya said. "We got five thousand e-mails, and we did not at all have the bandwidth to answer."

"If you sent us an e-mail, then I'm really sorry," Max said.

The solution, a member of the audience said, was to set up an online forum, and let the crowds answers their own questions.

What, someone asked, would it be like for the users? Ilya handled that one.

"Social networks are really good at acquaintances," he said, but not friendships, because of the difficulty in calibrating. "Privacy is not about not sharing at all; it's about being really clear who you're sharing with. We're thinking about different contexts of how people share information."

Diaspora would reflect an individual's social circles, and have clarity about which circles were getting information at a given moment. People ought to be able to control who they shared with.

"When you connect with someone new, you should not be, oh, here's an update, and all the embarrassing things you share with your closest friends you should not share with your coworkers. On the counter side of that, we want to facilitate sharing and make it really easy because that's what the Internet is about—sharing cool stuff with the people you care about."

Another question: Were they actually inventing new technology, or were they synthesizing existing code and putting a good interface on it?

Most of it, Rafi said, already existed—with one caveat. Discovery, or the ability to track down friends on existing networks, was a problem that had yet to be solved.

"User experience and user interface are inventions and that's the main part of the project," Rafi answered. "In terms of technology, it's basically all out there."

"We might not be inventing, but we might be the first implementers of it," Max said.

A final question was called from the audience: "So are you going to win?"

"Of course," Ilya said, breaking into a laugh, then adding a question of his own. "What is winning?"

The launch was mid-September.

# CHAPTER TEN

O
n the Wednesday after Labor Day, the four reassembled in San Francisco following their first sustained break of the summer. Max had flown to Linz, Austria, to speak at Ars Electronica, an annual festival for art, technology, and society that had been held in an old tobacco processing plant. The 2010 theme was "Repairing the World."

Ilya and Rafi sported new hairdos, stripes shaved along the sides of their heads. With Dan and Mike Sofaer, they had gone to Burning Man, the sprawling festival of utopianists held at the end of every summer in the Nevada desert, thousands of people camping, communing, and partying. They even staged a version of the Critical Mass bike rides that take place at the end of every month in cities around the world; at Burning Man, the ride was females only, with tops optional. Ilya thought it was an excellent idea, although he noted that the Burning Man version was not called Critical Mass.

"Sorry to miss you at Critical Tits," Ilya had texted to a friend.

On the long ride back from the Nevada desert to San Francisco, Mike spoke with them about the evolution of the project. In theory, he had been banished from business discussions early in the summer during the flare-up over acknowledging Mark Zuckerberg's donation, but he was Rafi's brother, and he had experience that all of them lacked. Plus they enjoyed going over to his house every few weeks for a night of board games. His involvement was delicate.

Now Mike asked: Do you want Max to be the CEO of Diaspora? They did not. He informed them that Max fully intended to be the CEO—in fact, he had told Mike that was his ambition on his first day in San Francisco—and that unless they acted, Max would be the top dog. Which of them wanted to take on the CEO job? Actually, none of the others. Back in June, when they had incorporated, Max had been listed, somewhat casually, as the president and CEO, though the others did not regard this as an actual decision to have him run the company. Besides, they were not thinking about the business: they were absorbed by building. Max was, too, but he was thinking already about how they would sustain and grow Diaspora.

If the summer had been full of sudden twists bringing them to places they had never expected to be, the fall had its own peculiar dynamic. It was the first time any of them had not been in school in September. As graduates, Dan and Max would, in any event, have been leaving behind the rituals of restarting classes, the familiarities and predictability of the academic calendar that had ruled nearly their entire lives.

They were in San Francisco, in the Pivotal offices on Market Street.

"My body feels like, okay, you're going to have a big change," Max said. "I think it's this residual muscle memory of going back."

"This would be the third day," Dan said.

The arrival of the school term churned Rafi's ambivalence about the whole enterprise. Not only was he the youngest of the four, but he was the one with the greatest emotional distance from the project. His personal compass was not pointed to computer science; his interests ranged across the law and politics, math and neuroscience. Learning how to hack was, of course, elemental to twenty-first-century literacy. But he had older brothers in the tech world. It was not his passion. He had been undecided about continuing with Diaspora beyond the summer; he had weighed returning to NYU for the fall semester, finally deciding in mid-August that he would join Ilya in taking a leave. "It's a little weird," he said. "There were classes I was looking forward to taking."

"What this means is that we are not just a summer project," Max said.

They had barely settled back at their computers when a man paused at their tables.

"Do you know where the Diaspora group is?" he asked.

Unscheduled visitors were part of their routine. Mostly, they were a class of gawker. Corporate clients of Pivotal would come to the sprawling space for meetings about their online business—people at companies like Best Buy, Groupon, Gowalla—and after their own affairs were dealt with, they would be escorted through the floor by their Pivotal manager. The office had props typical of the hip tech workspace: the pantry with unlimited treats and drinks, an archipelago of island meeting spaces, racks near the doors where scores of bicycles hung. It also had the Diaspora kids. Sometimes, visitors paused at their patch of table space. As this one did.

"That's us. How are you doing?" Ilya said. They always gave a friendly welcome, as they were temperamentally polite. Even if graciousness hadn't come naturally, brains did; they had come to understand, startling as it was, that they were celebrity geeks, and that these little walk-bys mattered to Pivotal, their generous host.

This particular visitor was a business graduate student.

"I was hoping to get an invitation to the developer release," the man said.

"Sure," Ilya said. "Are you a developer?"

"I'm just an average user," the man said.

He fell into a short conversation with Rafi, telling him that he did have some programming experience. Hearing this, Max braced himself to be asked for a job.

"So who are you seeing at Pivotal?" Max asked.

"I'm here to see you," the man said. "I've read a lot about the project and am very interested in what you come up with."

Whoa.

They deciphered the man's visit: he had come to see them. Unannounced. With no appointment. They knew the virtual eyes of the world were on them, but an actual person, a total stranger standing over their workspace, was a bit much. Suddenly the others could see why Rafi had been creeped out when someone tweeted that he had just left the subway.

"Leave your details," Max said. "We'll put you on the mailing list. The invites will be going out in the next week or so, if all goes according to plan."

This appeared to satisfy him. They watched him walk toward the elevator on the other side of the room. Then Ilya, grinning, crept up next to Rafi, and stroked his arm, as if he were fingering the hem of the garment of an emperor or rock star.

"Oh, Rafi, can I touch you?" Ilya teased. Rafi recoiled.

Their next plunge into the open waters of public life, a few weeks off, was very much on their minds.

In August, they had promised on their blog to release the developers code on September 15. This meant that their technical road map and the basic code foundation would be available on a public repository called GitHub. From that moment on, the global community of free-software developers could build on what the Diaspora team had started; anyone could suggest changes and submit them to the team, or anyone could just create a "fork" from their code and take it somewhere else.

The code release was big nerd news, but to the amazement of the guys, it quickly hopped the wall into a world expecting a fully formed social network, not what amounted to a wiring diagram. In draft form.

With dizzying speed, word spread in the nontech media that Diaspora itself, not just lines and lines of code, was going to launch on September 15. It was easy to understand how that happened. The blog posting had been their first public statement since the Kickstarter explosion in the spring. Their silence had been tactical, to avoid stoking speculation and stories about something that did not exist. As thrilling as the response to the Kickstarter appeal had been, they were acutely aware of the high expectations it had created, and they wanted to do nothing to accelerate them. Their intention for the summer had been to get work done, not give a running play-by-play of the line-by-line development.

Although the software sausage was still being made, many people assumed they would be sitting down to breakfast on September 15. Max wanted to update the post to make the situation clear; Ilya argued that they should say nothing more. Perhaps from his days in the high school drama club, lighting different parts of the stage, he had a keen sense of public spectacle. Sometimes what you didn't show was even more important than what was fully lighted. Another amplifying blog post, he argued, would dull the impact of their communications. The debate engaged them all.

"There could be a complete wave of haters," Rafi said.

That wouldn't be anything new, Dan said: "The beards are hating on us already—how come we didn't make it open source from the beginning?"

That is, some of the free-software purists were complaining that the Diaspora Four had not published their early scratchings. The notion that all drafting would go on in public seemed disconnected from the natural history of most human endeavors. They had been creating the framework of the project, and putting the fundamental elements of their vision into place. This is what would be made public next week.

Maybe, Ilya mused, they should have a special reservoir or holding tank for the bile that was sure to come their way. "We could make a ranters' list. Or a haters' list," he suggested.

The public's high, perhaps unrealistic expectations, were bundled into the whole Diaspora story, the great surge of support, the sharp hunger for a Facebook alternative. That same package included the shrinking of their private space, and an ill-formed conception by strangers that they were entitled to some ownership of their lives.

Some of it was intriguing. A friend had texted Dan to tell him that her summer Italian language class was reading a newspaper from Milan, and had spent an entire class discussing an article about Diaspora. Max's sister told him that a class at the Parsons School of Design had been assigned the task of designing what a distributed, privacy-aware social network should look like. In case anyone thought it was all dazzling, Rafi read aloud part of an online discussion about Diaspora.

"This guy sees us as a major threat, that we're a 'New World Order Mafia.'"

Rafi recited the rant: "'They're destroying our country and the rest of the world through their corrupt practices. Well, these are the same people who own Facebook, MySpace, Twitter, YouTube, Google and more. These people who make these sites gather information on people and monitor what we think. If and when we find out their attack on our lives, this Diaspora group is backed and financed by the same people. All of these guys are qualified to be Israelis, and they are going to do the same thing as Facebook.'"

Ilya cackled.

The post continued in the same vein, invoking 9/11 and people selling secret helicopters.

"I just want to give him a hug and tell him it's going to be okay," Dan said.

How long could they last? Dan was notoriously careful about how they spent the money—"He didn't want us to pay for Yosem's dinner," Max had complained—but they all shared a sense that they had to responsibly steward their funds. Moreover, their families were supplementing their Kickstarter reserves in various ways, like covering cell phone bills and trips home.

Over coffee, Max said that he was keen to hire engineering help—"heavy iron," he called it—but a few of the senior people at Pivotal had counseled otherwise. Instead, they suggested that Diaspora get a product manager, someone who would shepherd their existing work for release to the general public. Wouldn't it be hard for the group to have someone new calling the shots? Maybe they would not need an outsider, Max said.

"Pick one of us exclusively," Max said. "Our democracy of building a project is hard enough. You can't build a product with four cooks in the kitchen."

A friend from New York, Jamie Wilkinson, visited at lunchtime. He had recently written a piece of code that people could easily install on their browsers called the Google Alarm—a little monitor that sounded an alert anytime Google inserted tracking software onto the computer. "The idea is to make the invisible web visible," he explained. "You might have sixty little tracking bugs on a website. It's this totally quiet experience that nobody ever pays attention to."

The alarm made the bugs impossible to ignore. It also got him attention: suddenly, he found himself being called on as an expert on online privacy by television networks. He had also taught "Internet Famous" at a design school in which students were graded on their ability to generate online attention, no matter how spurious.

The Diaspora guys were eager to hear about Wilkinson's own sudden wave of fame, and its faintly ridiculous contagiousness.

"I did a CNN International interview about it and they asked me to come back and do a debate," Jamie said.

"Suddenly you're the go-to expert," Ilya said.

"It reminds me of Jonah Peretti," Max said.

"It's just like that," Jamie said.

Peretti, who had helped create the *Huffington Post* and BuzzFeed, had spotted an offer by Nike for customers to buy sneakers with their names or other message printed on them, saying the program was "about freedom to choose and freedom to express who you are." So Peretti submitted the custom ID "sweatshop" for a pair of running shoes, a poke at the company for the conditions under which its shoes were made. Peretti's request was turned down. His e-mail exchange with the Nike customer service department became a viral classic, and he found himself written about in, among other outlets, *Time* and the *Wall Street Journal*. He also appeared on the *Today* show to debate a Nike representative.

"He got all these invitations to go to worker-labor union talks," Max said. "He became like the expert on sweatshop labor. He thought it was funny."

With the developer release just a week off and the press already building, Jamie cautioned them that the inflated attention could damage the project, and to stick with their true identities.

"You just have to keep a thing, 'we're four kids from the ACM,'" Wilkinson said. It was sound advice. Regardless of the press frenzy, they had a long way to go with Diaspora.

"Even if we didn't change anything, we'd be here another six months," Max said.

"'It's the end of the beginning,'" Ilya said, quoting Winston Churchill.

A giant cartoon head—mostly a circle with a rascally grin and mischievous eyes—popped onto the Facebook news feed. It was the profile page for none other than the anti-apostle of Facebook, Ilya Zhitomirskiy.

"Just got an account! On a friending spree! Feel free to suggest some!"

Immediately, one of his friends posted the link to Mark Zuckerberg's Facebook page. Another, Fred Benenson, who had helped them navigate into Kickstarter, posted one of the first comments on Ilya's new Facebook page.

"Are you testing Diaspora? ;)"

Ilya replied that he wanted to understand the user interface issues that the others could speak fluently about but which he was unfamiliar with.

"Hence the Facebook account?" Benenson asked. "Or did you really cave?"

It was a matter of his own ignorance as a newbie, Ilya explained, known in geek land as a "noob."

"Found that I was asking Max really noob questions about how people communicate," Ilya said. "Also I'll probably use it for getting people to switch over when the time comes."

This was heartily endorsed by Jamie Wilkinson, who quickly weighed in.

"USE YOUR ENEMY'S TOOLS AGAINST THEM," he wrote.

They all seemed to be rolling with the stress, but that was an illusion. One evening, the four decided to see a film, *Scott Pilgrim vs. the World*, a comedy whelped from a graphic novel series. In the theater, the previews ate up ten or fifteen minutes, and Ilya could not sit still. When the feature began, he got up and left.

# CHAPTER ELEVEN

Time to go home. Dan had barely lifted his eyes from his screen all evening. To watch him, one might have thought that he was in the grip of some gravitational force that kept his head pinned in place. But really, it was a combination of his own absorption and disinclination to small talk. Every few weeks, I visited them, but kept out of their concentration tunnels. At some point, Dan stood, stretched, saw that I was seated a few tables away, tapping at the screen of a new tablet computer. The near scowl turned into a big smile. "Your own iPad? Sweet," he said. "Are you trying to write on it?" Indeed I was, foolishly.

In a moment, he had found and set up a wireless keyboard that liberated me from typing on the screen. Then he returned to his work.

Ilya and Rafi had spent hours at one of the standing desks, taking turns on the draft of a blog post that was supposed to explain what they were doing. Actually, they were doing precisely what they had promised: dropping a batch of software code into a public space on the Internet. But most of the readers of the post would be waiting to turn on their social network, and did not have the slightest notion that a mysterious ball of computer code was about to be made public, ready for kneading by other programmers. Explaining this was not simple. As ever, Ilya was keen to set the moment in a broader context, particularly the importance of free software; Rafi wanted to hone the message for the nongeeks. The two might have been negotiating the charter to a new nation.

They yielded every half hour or so to Max, who hovered. Clustered within a few yards, they had been a solitary late-day encampment in the Pivotal offices, a nearly city-block length of uninterrupted space.

Ilya wandered toward Dan, who was fiddling with sections of code that would polish the look of the site.

"How is it going?" Ilya asked.

"We're only going to get one chance to do this," Dan said.

That morning, Eben Moglen told Rafi that he would be invited to testify at a congressional committee that was holding hearings on Internet privacy. With the press of work for the next code drop, the invitation had dropped off his radar. Rolling kinks from his neck, Rafi absently stroked his head, which was freshly shorn to the scalp.

Who had done his hair?

"Walgreens," he said. "You can get a haircutting kit for thirty dollars."

The styling, it turned out, was one of Ilya's do-it-yourself projects. A few weeks after the Burning Man festival, their pictures were being taken for a feature in *New York* magazine, and the stripes Ilya and Rafi had cut into their hair before going to the festival had long since lost any semblance of order. Even at age nineteen, Rafi had plenty of sense about testifying in front of Congress—his father, the federal judge and State Department counsel, knew all about the process and perils—but he did not have the years to carry off a cue-ball skull.

"It'll grow back," Ilya said, confidently drawing on DIY adventures in amateur barbering that went back to high school.

Around ten, they all agreed that they had done everything they could for the day. "We'll come in in the morning, give it one last check, and push the button at eleven," Dan said. "People are waiting for it."

He jumped on the subway and rode out to the Mission, where he had recently found an apartment, leaving only Ilya still bunking in the flophouse. Dan had not gotten around to furnishing his new space. Bedless, his sleeping arrangements consisted of a few pillows and a blanket. It was moot that night. Around midnight, he was still fingering bits of code. He tweeted:

sleep? What's that?

————

So consumed were the Diaspora guys with the pending release, none of them noticed a tech story ricocheting around the Internet that night. They were not the first incarnation of the heroic geek, carried on the shoulders of digital acclaim. Working a few blocks away in San Francisco, a man named Austin Heap had been toiling for more than a year on another Internet project that had captivated people with the transformative, liberating possibilities of the online world. He was twenty-five.

During the June 2009 Green Revolution in Iran, Heap had watched as the regime filtered Internet communications, identified dissidents from their digital footprints, and heaved them into prison. A software developer in San Francisco, Heap had little interest in politics. The crushing of the uprising in Iran awoke him. What the activists needed was unfiltered access to the Internet, and encrypted communications, so that they would have a fighting chance to carry out their work without being instantly identified, thwarted, and beaten to a pulp.

Heap said he would build that precise tool. It would be called Haystack, and activists who used it would be like needles hiding in it. Teaming up with another developer, he created the Censorship Research Center. Donations poured in. The new U.S. administration of Barack Obama, committed to the use of soft diplomatic power, backed the development in small but vital ways. For instance, Haystack ran into none of the hassles that faced the first developers of public encryption, who had been legally barred from sharing their work until Eben Moglen and others fought federal prosecutors in court and elsewhere. Even so, it remained exceedingly difficult to export encryption technology, which was regulated by provisions of a Cold War–era law written to forbid Americans from sending nuclear weapons technology abroad. Haystack faced no such problems. It received vital licenses from the U.S. Department of State, the Commerce Department, and the Treasury.

At the moment a totalitarian government was applying digital screws to its people, Haystack and Heap brought the promise of salvation. He was hailed in an HBO documentary. That summer of 2010, on the anniversary of the Iranian uprising, *Newsweek* published a glowing profile. The *Guardian* honored Heap as "Innovator of the Year" for protecting free speech. Crisis is bound to hero by the covalent bond of narrative. Iran and Heap. Facebook and Diaspora.

In short, if Diaspora had an avatar, they would not have had to look beyond Austin Heap and his Haystack, subject of celebratory, uncritical press accounts and recipient of popular support that was expressed, concretely, with cash donations.

Like the Diaspora guys, Heap forecast that his project would be ready to go after a summer or a few months of intensive coding. He was wildly optimistic about how quickly he could build the software. Meanwhile, the acclaim mounted; good intentions were confused with the actual goods. Behind the scenes, security and cryptography mavens urged Heap to share the work in progress. He said that was too risky. This seemed like common sense to people who knew nothing about cryptography, but was the very definition of folly to experts. The design principles for secure military cyphers were laid out in the nineteenth century by Auguste Kerckhoffs, a Dutch linguist and cryptographer who advocated that "the system should not depend on secrecy, and it should be able to fall into the enemy's hands without disadvantage." It was probable, if not inevitable, that an enemy could figure out exactly how a system worked, but what would protect the message were the unique keys, shared secretly by the sender and the recipient. Without them, the message would remain indecipherable, no matter how much of the design the enemy knew. A single deciphering key set could be captured, but only its accompanying message would be compromised. Every additional secret was another weak spot, another likely point of failure. Security experts on a private mailing list that covered "circumvention and obfuscation" were making a ruckus about the secretive development of Haystack, and Heap was publicly challenged to let experts review the code by Evgeny Morozov, a Belarusian scholar at Stanford University who wrote frequently on the relationship between the Internet and political dissent.

Instead, by the end of the summer of 2010, Heap arranged to hand deliver CDs of Haystack to Iranian dissidents for "stress testing." Inevitably, a copy made its way to security experts, and one of them, Jacob Appelbaum, demolished it within a few hours. Not only did it not protect the users, it did precisely the opposite: users of Haystack would leave a trail of digital bread crumbs that led directly back to them. The chief developer working with Heap quit. A day later, Heap announced that he was ending the testing of Haystack until the problems could be addressed.

The story hit the *New York Times* website on September 14. It was a lesson, commentators said, in the hazards of media cheerleading uninformed by the slightest understanding of what was technically feasible.

The following morning, the Diaspora Four were at their tables before most of the Pivots had poured their first cups of fair trade organic coffee. The night before, they had managed to convince themselves that they had very little to do to keep the promise of the September 15 release. Three chores, really. First, they needed to have a tweet poised to launch at the moment the code repository was opened, directing the community of backers to the *JoinDiaspora* blog for an update that would elaborate beyond the 140-character cage of Twitter. The second was the blog post, letting their nontech supporters know that they had reached this milestone, distant though it might have been from the ready-to-roll tool that the donors had thought would take the guys just the summer to write. Finally, they needed to put the code through its paces again, to make sure everything was running properly, or near enough. It did not have to be perfect; indeed, the very reason for posting its entrails was to have crowds rise in their wisdom to fix and improve what the Diaspora Four had started, the long tail of the Internet working its magic.

Still, they didn't want to be public laughingstocks, which they immediately discovered was a distinct likelihood if other developers were to run their code in its current state. A bug had crept in. Somehow, Diaspora was not allowing users to add friends: nominally, a social network, but one that offered only solitude.

"It's a show-stopping bug," Max said.

They crunched backward through the code, trying to find what had gone wrong. Before they knew it, noon arrived. Pivotal Labs was holding its customary Wednesday talk, but no one from the Diaspora crew even glanced toward it. By the time they wandered near the buffet, they found a few spoonfuls of couscous and a half-plate worth of grilled vegetables. So potato chips for lunch. Meanwhile, the bug was not yielding. In cyberia, the natives were getting restless. They posted complaints on

Facebook, which had been hosting a Diaspora group page from the beginning of the project.

"Vienna is waiting for you . . ." Helmut Kreutner wrote at 12:35 P.M.

In Argentina, Federico Carrizo noted that it was heading for 2:00 P.M. "Still waiting," he wrote.

A man following the Diaspora discussion in upstate New York invoked a website whose only function was to repeat the same welcome announcement and introductory remarks that ran in a maddening loop.

"WELCOME to ZOMBO COM," wrote Andrew Charles Cayer.

For people waiting in Europe, September 15 was close to having come and gone. At one in the afternoon in San Francisco, it was ten P.M. in Paris. From France, Jean-Baptiste Tobé Berlioz demanded: "where is the source code? give me the source code!" In Cagliaria, Italy, Antonio Lorenzo Picciau wrote, "come onnnnnnnnnnn."

There was debate, bawdy at times, over what was taking so long. "Either someone overslept and forgot to do all what they said, or it's vaporware and they're spending that 200k!" suggested Jason.

"i think they are drunk and stoned somewhere in las vegas, married with whores," Mike Vourk wrote.

"The whole thing was on Ilya's laptop's hard drive, and they spilled beer on it last night," Kyle Bennett suggested. "They've been up all this time frantically typing it back in from printouts."

"Guys, could we get just some idea of whether to keep watching or just come back at a certain time? It would free up my day, and more to the point, my 'refresh button,'" pleaded Chris Larson.

Dan, who had deleted his Facebook account months earlier, was able to ignore the online demands for the code. He did nothing else until he had tracked down the bug and got the program to run again. The blog post was another matter. Ilya was not satisfied that a dozen drafts and false starts had rendered it into plain language. Max gnawed at his fingernails.

Rafi dipped into the Facebook page, and saw kindred spirits, suspicious of hype. If he were sitting with them, he would have had precisely the same perspective. In fact, he almost had it and wasn't sitting with them.

"This is awesome—these people are going nuts," he said, reeling off a few comments, concluding with one that he deemed worthy of the Facebook seal of approval.

"'I'm sick and tired of waiting. R u joking with us?'" Rafi read, then announced: "I'm going to 'like' this one."

There was more, and only so much cynicism that they could laugh off. "I'm not surprised that it's not out yet ... i doubt it will even come out today. They don't have anything done and they're probably hoping we fix it/build it for them," Joseph Pereira said.

Jean-Baptiste Tobé Berlioz tried to calm him. "C'mon," he wrote. "It's not like the world will end today." The cybergathering was one of developers—amateurs in the original Latinate sense: people who worked or played or created or built out of love for the task. They spoke about the coding language that had been used—Ruby—and discussed learning how to use it. They kept track of the people signed up for the Diaspora repositories on GitHub, a powerful tool for collaboration among software developers.

"It's midnight in France," wrote Jean-Baptiste. "I'm going to sleep now (drunk already). I hope to get the source code tomorrow."

Then he emended: "Ok, ok, last glass of wine."

"Dear Diary," responded Mike Vourk. "It's 1:18 in Greece, no wine in the house (beer, it's gooood for you.)"

Max and Dan had scrubbed through the code, but Rafi still wasn't sure. It was possible to lease time on simulation machines to run tests of the software, but they had not gotten anywhere near that stage.

"We still haven't verified that it works," Rafi said.

"Yeah, it works," Max said.

Rafi tried a feature they had available for their own in-house tests.

"Zombie friends is not working," he said.

"No one is going to care," Max said, a touch dismissively.

Dan assured him. "Those sorts things that are, like, internal."

Rafi was mollified. "All right, just to clarify that it works. Let's deploy. There is a strong possibility that it's functional."

Max looked up from the page of plaintive and angry posts on Facebook. "We've tortured these people enough," he said.

That meant Ilya had to finally let go of the draft he was working on for the blog post. "All right, dudes, how does this look to you?" he called out.

Dan leaned over his shoulder and read the draft, which began: "Today, we are releasing the source code for Diaspora. This is now a community project and development is open to anyone with the technical expertise who shares the vision of a social network that puts users in control."

Ilya had included screen shots that showed what the working version of Diaspora looked like. Dan dragged them into a new arrangement on the screen. Max had logged into their Kickstarter page, so that word would get to everyone who had donated and subscribed. He typed two words—"It's here"—that would be sent to Kickstarter as soon as the code was made public.

"Let's get everything ready and then have a moment of silence," Max said.

"Then get slaughtered," Dan said, grinning.

A sense of high occasion replaced the unspoken urgency of the afternoon, but now that they were ready to push the button, they were missing one person.

"Where's Rafi?" Ilya asked.

"Probably at the gym," Dan said.

A moment later, Rafi returned from the vicinity of the men's room, and Ilya skipped up to him, his hands up in the air for a high five. Rafi fended him off with a forearm.

Max thought through the orchestration. The announcement was being broadcast through several venues on the web, so everything would have to get pushed out within thirty seconds—opening the repository on GitHub, posting the announcement on the blog, sending out the tweet, and blowing the trumpets on Kickstarter.

"Let's all hold hands!" Ilya said, laughing, a bit embarrassed because he meant it.

Actually, their hands were otherwise occupied. Ilya pushed one send, Dan another, Max a third. "It's pulling," Dan said.

Ilya spotted the draft tweet, which said, "Developer release of Diaspora." This felt banal.

"We should make it more epic. 'One small step,'" he said, grandly invoking the words spoken by the astronaut Neil Armstrong at the moment he became the first person to set foot on the moon.

"It's a tweet, man," Dan said, a bit exhausted. He and Ilya jousted over the language for a moment.

"Hurry up!" Max called. "Someone else is going to tweet our shit."

Dan typed, "Diaspora is real." Then he stretched his arms overhead. "Finally," he said.

Rafi spoke. "I'm starting to get hungry," he said, and the others realized that they had barely had a bite to eat all day.

Still, they were gripped by the numbers of people clicking onto the code. In the first minute, 349. "We're on *Hacker News*," Max said triumphantly.

Ilya's phone was ringing and would not stop. Finally, he looked at the screen and saw that it was Jamie Wilkinson, their friend and ardent champion.

"Yo, bud, we're out there!" Ilya said.

He paused for a moment.

"Typos on the blog post?" Ilya said. "Are they worth fixing? I'm too high on adrenaline and all this other stuff."

The error, it turned out, was not a small one. They had spelled Diaspora as "Disapora."

Max fixed the spelling. Their first troll filed a comment on the Internet relay chat, and Rafi read out his message in a monotone: "'I want to stick my dick in this code. But it's coated in herpes spores.'"

They roared and high-fived one another.

Four baskets of bread vanished at the steak house before the waitress took orders. It was their first food of the day. Someone mentioned Evan Korth. Ilya slapped his forehead. "I promised to send an e-mail before we pushed the button, but I totally forgot," he said.

Dan reminded them that after returning from Burning Man with them, Korth had left his belongings at the Sofaer home in Palo Alto.

"He's sleeping in his office naked," Dan said.

Ilya phoned Korth in New York.

"So sorry," he said. "But your name is all over the Internet." "Korth" had been the password for the crash-test dummy they had used during development, and they had left it in the code.

Mike Sofaer had joined the dinner celebration, wearing a T-shirt printed with the title "National Sarcasm Society." Below that, it said: "Like we need your support."

The Sofaer brothers were in fine form. "Filet mignon," Rafi told the waitress. "Very rare."

"I want the filet mignon, too," Mike said, "but I want it rarer than his."

Mike's phone could be turned into a portable router, so they all were able to get Wi-Fi at the table on their phones. This allowed them to read out responses from Facebook and the blogs. Already, they had four hundred "watchers," people who had signed into the code with the ability to propose changes. If just 10 percent made contributions, they would be getting rapid improvement.

"When we want something done, we can go away," Dan said.

"We can take naps," Rafi said.

Thrilled as they were at the prospect of having other people join them to write and rewrite the code, this meant new legal issues, which seemed technical but were at the ethical core of their undertaking. The license in a free- or open-software project is the governing document of the collaboration, spelling out what contributors of code could expect. Typically, they had to assign their copyrights to the project. Dozens of types of free licenses exist, but under the core principles, developed by Richard A. Stallman and Eben Moglen, no one could stop anyone else from copying, modifying, or selling the code. That lack of restriction never went away, no matter how many times the program was modified; through all its iterations, every time it was improved or updated, the intellectual property remained fully available for inspection or use. Other licensing arrangements permitted some retention of rights by the people who contributed the code.

The choice of license included practical concerns of the community of specialist coders who would be working on the project. The Diaspora guys gave the license almost no attention until it was time to make the code public. At the last minute, Rafi scrambled to get one, and the license that went up with the code was not entirely "free" under the

Moglen-Stallman regime. The code contributors retained some rights. So did Diaspora Inc., the for-profit corporation they had formed in June.

At dinner, Ilya said they needed to revise the terms; he felt that they did not fully reflect the principles that the project was meant to embody.

"You should give it some thought," Mike Sofaer counseled.

"I've given it a lot of thought," Ilya said.

The license was broken. Fixing it was a task for another day.

After dessert, the group stepped into the brisk San Francisco night and walked toward Market Street. I fell in with Rafi and we chatted about Moglen's proposal that Rafi give testimony as an expert on the subject of Internet tracking. The Sofaer family knew the ways of Washington.

As we walked along Kearney Street, I asked him what he would tell Congress.

"The truth," Rafi said. "If there's one thing our father taught us, it's tell the truth to Congress and judges."

We switched our conversation into the cross-examination mode of congressional hearing rooms.

"Isn't it true, Mr. Sofaer, that you created this 'Diaspora' for the sole purpose of getting girls?"

Rafi did not hesitate.

"With all due respect, Mr. Congressman," he said. "Is there any other reason?"

Within hours of the code's release, other hackers identified security vulnerabilities. "The bottom line," said one prominent critic, Patrick McKenzie, "is currently there is nothing you cannot do to someone's Diaspora account, absolutely nothing." McKenzie was interviewed for an online computer trade publication, and his comment was repeated in hundreds of publications.

The troll commentariat rose in unison to declare the project dead on arrival. "Diaspora, security catastrophe," dominated the narrative. The guys were not alarmed, but people who cared about them were. Carolyn Grippi, following the coverage online, called her son one morning in a panic. "Mom, calm down," Dan told her. "It's going to be fine. Most of it is already fixed."

Indeed, most of the security problems were quickly patched up; the very purpose of releasing the code publicly was to let hackers and coders bang away at it and find problems before ordinary people started using it. The failure to do this had led to the instantaneous demise of Haystack, the prematurely acclaimed digital lifeline offered to Iranian dissidents by Austin Heap. Writing in a blog later, McKenzie reported that he had found "a half dozen critical errors" in the code they had launched. Several variations of the errors let users log in as themselves, and then mess around with the data of other people, or even take over their accounts. Another vulnerability was the encryption that Diaspora encased around the communications between two users. The code permitted an attacker to overwrite the existing encryption key and capture someone else's private messages. McKenzie spelled out the flaws only after they had been repaired.

Commentators on McKenzie's blog debated their significance. Some insisted that they showed incurable naïveté on the part of the Diaspora team; others said such early mistakes were common when code was developed with the language and framework used by Diaspora, called Ruby on Rails. Most of those projects were backed with venture capital, which did not do wholesale public code dumps during the building. Naturally, Diaspora looked worse: the guys had simply pulled up the shades on what was normally a hidden process. "It's just that we don't get to see the code of most venture-backed start-ups at launch," wrote one reader of McKenzie's blog.

Max, Ilya, Dan, and Rafi had already moved on to other refinements, though the "security problems" of Diaspora were frequently invoked, if rarely spelled out. The reality of the process, rather than the noise around it, was consuming them.

"Lead, follow, or get out of the way," Ilya said. "The armchair quarterbacks. Security stuff, we fixed most of it right away. I think we are way more thicker-skinned than we were at the start of it."

With relish, he quoted a tweet from a hacker who was following the project: "'Now that Diaspora is out, everyone who complained that it is vaporware is going to complain about the way it works.'"

# CHAPTER TWELVE

After four months in a city where they knew no one but one another, isolated by the drudgery of waking hours spent at Pivotal or hunched over laptops in their dingy quarters, Ilya, Dan, and Rafi craved a change. They came back to New York at the end of September on various pretexts—an open-source video conference sponsored by Mozilla where Ilya and Max were on a panel; a lecture that Dan was giving at a design conference in Providence, Rhode Island; and consultations with some lawyers for Rafi about licensing arrangements for coding contributions to the project. But really, three of them just needed the ordinary fruit of college life: a network of friends with whom they could hang out. In person. Not online. No less than the others, Max also was conscious about finding balance in his life, but he saw no need to go back to New York for it; in his mind, he had definitely closed the East Coast chapter and was fully committed to the life of a budding entrepreneur in the Bay Area. Nevertheless, he, too, had come back east for this break.

The people they had been spending their days with at the Pivotal offices in San Francisco were almost all professionals hitting top stride in their careers; that is, they were older, settled into family and living arrangements.

New York, on the other hand, teemed with friends and peers.

"That's why we're slightly heartbroken about this epic city," Ilya said over coffee near NYU.

Being back in New York spun the flywheel of his mind ever faster. He had stopped by an anarchist collective, ABC No Rio, which housed a group called Food Not Bombs. "One thought, to add to my list of projects when I sort of run out of projects, is to make penny stoves," he said, describing a contraption made from soda cans and using ethanol as fuel. "You could make lots of them and give them to homeless people. There's so much food that goes to waste."

Ilya paused. "I was much more interested in this last summer," he said. "I get really excited about projects and, like, one-tenth of them really happen. Because there's never enough time for everything."

Musing about San Francisco, he reported that he had managed to find someone with whom to home-brew beer. Something to do that did not involve coding all day, work he could do with his hands, and that had an end. The deluge of criticism about the security holes, which had been fixed relatively simply and quickly, was an example of nonproductive carping.

"There are so many things simply lost to passivity. Instead of just complaining about something, just do it. I think that's very much the way things get done. Being passive and cynical is just boring –you have these awesome movements that are happening."

He reeled off a list of developments that he saw as positive: the rise of hacker spaces in major cities; the emergence of copyright licenses that liberalized the use of creative material; the legalization of jailbreaking an iPhone—meaning that an owner could load it with applications that were not approved by Apple, which might make the device vulnerable to malicious hacking but also created easier paths to innovation. And through Diaspora, he and his friends had shown the potency of crowdsource funding for projects on Kickstarter and other platforms like it. "That completely undermines the way funds are raised for creative projects," Ilya said.

On his pilgrimage of encounters in his junior year of high school, he had reckoned that about one in four people were alive to what was going on in the world. "This is something I think about a lot," he said. "There are people who are completely unaware of hacking, making and creating things just for the awesomeness. Then there are people who are aware and regret that they don't do enough projects. And then there are people who are actively working on projects.

"An epic project for the future would be to somehow use social pressure to move both of these groups in the area. It's like you're not aware of awesome people doing awesome stuff. You should at least become aware of it. We need to move both of these groups in the right mental direction.

"It doesn't matter what it is. Whether it is basket weaving. Or beer making. The world is run by those who show up."

Text alerts pinged like hail from Rafi's phone. He was delighted. His NYU friends were meeting, and they were sending him word because they knew he was around. As it happened, he was on the third floor of the Courant Institute, in the computer room at NYU where he and the other three guys had taken the world by storm six months earlier.

He made a call.

"Lots of people are there? Great," he said. "I'll be over."

He listened for a moment. A scowl crossed his face, an unfamiliar cloud. It seemed that the bouncer at the bar was checking ID cards. Rafi had celebrated his twentieth birthday not long before, still a year too young to drink. Plus, he looked sixteen.

"Then I guess not," he said into the phone.

Slight bummer, but not a calamity. He was just glad to be back in New York. Diaspora was a great opportunity for him, and he would stay with the project for another year at the most, or maybe just a few more months, until the end of the second semester. But he would have preferred, as Ilya and Dan did, that they were working on the East Coast. He had a level-headed view of the situation.

On the one hand, no one in the group had a better handle than Rafi on the nuts-and-bolts support Diaspora was getting from software developers around the world, giving their own time. He was managing the suggestions and proposed revisions that were coming in every day, nearly every hour. There had been five major submissions. And every day, it seemed, someone was translating Diaspora's pages into a new language.

Yet he was cautious about whether this enthusiasm would amount to anything. The translations would soon be obsolete if they weren't already, because the code was being revised so often. Plus the general public did not care about the intricacies of software development, and whether it was done with purely altruistic motives.

"For consumer software, usability is way more important," he said. "It is not an impressive piece of software. None of the four of us are genius engineers."

An e-mail popped onto his screen, and reading it, he made a face. This whole business of him testifying in front of Congress was getting out of hand. A staff member on the committee wanted seventy-five hard copies of his testimony ahead of time. "This can't be serious," he said. "Don't they have printers?"

Sure, the idea of testifying before Congress was exciting, but he had too high a regard for what he did not know to feel comfortable sitting as a witness before a committee. The last thing Diaspora needed was more hype. Plus they would undoubtedly dream up new hassles. He was bailing out.

The hallway at the Fashion Institute of Technology teemed with people, but a woman walking along spotted a familiar face.

"Max! Congratulations on the article! You're famous," she said.

"Thanks a lot," Max said. *New York* magazine had run a lengthy, insightful profile on the four, but included it in a special section entitled "Who Runs New York?"

"It's ridiculous," Max noted. "I don't even live in New York anymore."

That weekend in October 2010, by far the coolest place in New York for a young, hip tech person was a basement at the Fashion Institute of Technology, a campus of the State University of New York. Mozilla was hosting a conference on new web standards for video. In defense of independent filmmakers and videographers, an army of young techies had gathered to make sure the web did not turn into a vending machine for Hollywood studios and television networks.

The particulars held little interest for Max, who essentially was along for the break with Ilya, Rafi, and Dan. The others wanted to move back to New York to continue their work, but Max was adamantly opposed. "New York is so over," he said. "I'm done with it."

More than a physical place, he had migrated to the land of entrepreneurs. To keep moving forward, he had become certain, they needed financing to hire more technical firepower. He and Dan had some experience in programming, but Rafi and Ilya had none. In particular, Max was distressed by the quality of Ilya's code. Moreover, the team ate up hours quibbling. Someone from Pivotal with authority and experience could cut those debates short. "We need to hire a Pivot," Max said.

Another familiar passerby spotted Max: Tantek Çelik, a member of the federated web brigade that had met in Portland.

"How's it going?" Tantek asked.

"I'm still standing," Max replied.

From the perspective of Tantek and the others in the federated movement, Diaspora arrived at a critical moment. Not only was Facebook chronically overreaching, but the microformats built by Tantek and the others were now in shape to be used widely for federation. That Diaspora, with all its attention, might be the one to put them together and make the breakthrough to a broad audience did not trouble Tantek.

"We finally have the building blocks to build out this social web," Tantek said. "It really feels like we have been empowered with all that has gone on with you guys."

"We're definitely pushing as hard as we can," Max promised, but suggestions for fixes, "patches" to the codes submitted by volunteers, were coming in so fast they were inundated. It was an embarrassment of riches.

"We had been the four dudes just working it out, but at this point, we're totally overwhelmed," Max said.

To Tantek, the stakes went beyond the logistics of one group of kids struggling with attention and one another. Using the web was like walking on soft grass, leaving traces of every step. For the indie web troops, the stakes were control over what footprint you leave in the world.

Tantek had some advice on getting clarity on what needed to be done first: eat the dog food, reiterating what he had said in Portland.

"Running it on your own site will prioritize that stuff so fast," Tantek said. "The best programming comes when an individual has an itch, and they do whatever it takes to scratch it."

Max was a heavy, heavy user of Facebook, but not of Diaspora. It was a running joke in the group that while building Diaspora, he was still checking his Facebook page every two minutes.

But no dog food, as yet. In fact, he had not yet taken Diaspora out for a real-life drive. "I'm interested to start using it," Max said. "And Dan is, especially. Dan is the one who doesn't have a Facebook account. He just feels totally isolated from people. He's desperate."

"And he's resisted getting an account?" Tantek said, impressed.

No, actually. Dan had had one, but he vaporized it.

Dan spent a weekend in Providence, Rhode Island, about two hundred miles north of New York; he was invited to lead a panel at a conference called "A Better World by Design." He and the audience, which included many students from Brown University and the Rhode Island School of Design, talked about online relationships and what they had to do with real life.

As the session wound down, a young man raised his hand.

"Do you have a business plan?" he asked.

"We don't have a business plan," Dan said. "We just want to make software."

Break over, they returned to San Francisco for an October of humming days and nights.

One evening, the scent of hot pizza, still steaming from the oven, set off a small-scale stampede toward the Pivotal kitchen. Floating through the windows, a raucous din rose from Market Street. It was a time of day when both the streets of downtown San Francisco and the offices of Pivotal software were normally empty or close to it. But on that Thursday night near the end of October, the San Francisco Giants were playing the second game of the 2010 World Series. It was far more fun to take in the game with company, even strangers, in bars, or, if in a car, to lean on the horn.

At Pivotal, the meeting area was filled with people paying no attention to the baseball game at all, but who wanted to help Diaspora. Among them was a professional software developer named Sarah Mei, who had convened the gathering.

Even as clouds of venomous criticism of Diaspora blew across the Internet, an ant colony of people had started working on it from all over the world. Most of the time, they could be seen only in trails of terse comments left on GitHub, the repository, or in Internet chat rooms. Sarah Mei had tightened the code immediately after its release in September, speeding it up. Ilya and she struck up an online conversation.

"I actually work at Pivotal," she said.

"Are you in San Francisco?" Ilya asked.

She was.

Ilya digested that for a moment. She had been down the hall the whole summer, not on the other side of the world.

"We should get lunch," he said.

She had attended the presentation they'd given in the summer, and had taken a hard look at what they had done. First, she was impressed with how undefensive they were about criticism. They did not seem capable of getting their backs up: a decidedly unmacho and—she thought—smart way of handling things.

Second, the code was in much better shape than she expected. So she began to pitch in, quietly, without introducing her physical self to them. Though she was still working for Pivotal, her regular contributions to Diaspora made Max, in particular, happy. Hers was an experienced and authoritative voice, precisely what they needed. He wished he could hire her, but at Pivotal's rate of fifteen hundred dollars per day, she would burn up their funds in a few months.

In October, Sarah announced plans to hold a Meetup, "Contributing to Open Source 101 with Diaspora." Meetups were just the kinds of live, in-person gatherings that were never so easy before people were online all the time, supposedly losing their ability for normal human discourse. This one was completely booked within a few days. Organized through a social website called, not surprisingly, Meetup.com, which had been created in 2002 by Scott Heiferman, Meetups propelled the candidacy of Howard Dean for the Democratic presidential nomination in 2004. They had many purposes beyond politics. People enrolled to be alerted about get-togethers on subjects they were interested in—weight loss, politics, fishing, knitting, hiking, technology, and thousands of others. It was an excellent tool for Diaspora; groups interested in it held Meetups in about a dozen cities around the world. Their friend from Stanford, Yosem Companys, had been a big advocate of getting them going. The one in San Francisco would prove to be among the most active.

Rafi was at the podium.

What was their next goal for release?

"We have a goal to release something—an alpha, really—before Thanksgiving," Rafi said.

Rafi explained that they had a website, bugs.diaspora.com, where people could let them know what needed to be fixed. With slightly more than a month of responses from all over the world, though, he asked his audience not to be too quick to sound the alarm.

"When you find a bug and you want to submit it to a bug tracker, the worst thing to do is to immediately put down 'New Bug,'" he said, because they were getting multiple reports on the same problems. "A lot of time is going to be sunk into figuring out, no, we don't have 220 bugs, we actually only have 50 bugs, because really there's one incredibly aggravating bug that 30 people have submitted."

Then a man sitting near the front spoke up. His name was Sascha Faun Winter, a design and software consultant, and he did not pussyfoot around the main public critique of Diaspora.

"Have you guys addressed publicly, or otherwise, the security concerns?" he asked.

A small smile seemed to live perpetually on Rafi's face, a look of bemusement. It did not fade even as he took on this most basic, and possibly fatal, challenge to Diaspora.

"When we first did our release, there were a couple—really, just one—blog posts about security holes," Rafi began, giving background to the rest of the crowd in the event that they had not seen the criticisms before he directly answered Winter's question.

"Basically, what we did with that was we fixed most of them—I mean, we already have security warnings on the site and put a bunch more on—and then put the rest of the holes on our bug tracker and have been fixing them as we went along."

The four dudes shared a striking quality of poise in public, even when faced with criticism that felt excessive. Rafi's face was the picture of calm. It was part of why Sarah Mei was willing to take time out of her life for their project.

Rafi reached into the core of the critique: "The assertion that it would be impossible for us to build a secure application is just—sort of, sort of bizarre. Maybe it would have been impossible for us if we were locked into a room, trying to get security skills out of textbooks or something. But we're not. We've had a lot of help from security experts. We've had a lot of help from experienced rails developers."

Another member of the audience broke in to scoff at the premise of the critique.

"Is it possible to create a totally secure app? Nobody's built it. That's an unreasonable expectation. What should be reasonable is people bring up security problems that the community can work on. That's how UNIX got built."

Before the group started digging into the code, there was one more question about why they were going ahead with an alpha release in a few weeks. Was the goal, someone asked, that people would use it?

"The goal is that people would test it," Rafi said.

He paused for a moment, looking over his audience, the small smile never leaving his lips.

"Nontechnical people."

After five months in the boarding flophouse, Ilya thought that just about any decent, sociable place would look good. As he toured the three-story house at 757 Treat Avenue in the Mission District at the end of October, every floor was buzzing with young people. They were cooking on grills, playing pianos, talking earnestly. His tour guide was Gardner Bickford, a tall, blond, bearded man who was the nominal captain of the apartment on the third floor, and possessed of the classic manner of the laid-back Californian. Gardner had posted a notice on Craigslist, but he wasn't sure what to expect; he had just spent two years in New Zealand and was unfamiliar with the savage competition for rentals in San Francisco that followed the bloom of young tech people with money in their pockets, desperate for a place to stay. The first two people answering the ad arrived at the same time.

A little while later, Ilya arrived, and Gardner showed him around. They climbed back up to the third floor.

"What do you think?" Gardner asked.

Ilya didn't hesitate. It looked like paradise. It even had a name: the Hive.

"I'll take it," he said.

A few years earlier, that same building had toppled into one of the holes in the urban world created by crack cocaine. Then it was taken over by a kind of digital collective, Couchsurfing.com, a place for young people on the move to swap digs. Actual glass windows replaced the plywood boards, the alleyways were swept clean of the glass vials that had crunched beneath every footstep, and the place was restored not only to a state of good repair but to one of ample possibility. It shared a sprawling concrete backyard with the adjoining house. There was even a permanent tent/marquee set up in part of the yard, an outdoorish retreat for the damp San Francisco nights. The Couchsurfing people lived and worked in it, including Cory Fenton, the computer programmer who had dreamed up the project. Then Gardner, who had done volunteer coding for the site, was moving back to California, and the couch surfers needed someone to take over the lease. He stepped in. Not long after, so did Ilya. At last, he lived in a place with people who interested him. It was full of promise.

That week, Ilya wrote a post for the project's official blog. He listed the improvements they had made in the six weeks since the code had been made public. It was easier to send messages out, to post public messages not only to the Diaspora site but also to Facebook and Twitter, and the whole "getting started" experience had been made friendlier.

"Our basic feature set is almost done," he wrote. "Once that is stable, we'll set up an alpha server so that anyone, not just developers, can try Diaspora and help us improve it. We're shooting to do this before Thanksgiving."

They were off to the races—or so it seemed.

One evening, Ilya walked out with Mike Sofaer, who was on his way to train in a gym two doors away. He stood in the gym lobby and poured out his heart.

Ilya was overwhelmed by the work. He was no good. He did not have the coding skills the others had. Mike was about the only older person he knew well enough in San Francisco to trust.

"You don't have to be the best programmer in the organization," Mike said. "Not everyone can be the best programmer."

Ilya fretted that he could not do the high-level architectural design as well as Rafi.

"That doesn't matter," Mike said. "It's not important that you don't do it as well as him."

What mattered, Mike said, was his drive and enthusiasm, things he communicated better than anyone. His spirit. "You're the heart and soul of it," he said.

"Thanks, man," Ilya said.

Mike didn't think he had persuaded him.

# CHAPTER THIRTEEN

The days and nights before Thanksgiving 2010 blurred, an all-out sprint. They had kept the first promise, to release code for other developers by the end of the summer. Facing a deadline for its debut outside that rarefied world, they converted the scoffing nastiness of many critics into high-octane fuel, and stayed up all hours chasing flaws. They would show people, not tell them. That meant paying attention to important bugs that were coming their way from all points on the globe. The nature of the problems could be trivial typographical errors, or misstated function names; they could be errors in logic or security. As the number of people using a piece of software grows, more bugs are found. Virtually all software—whether apps for smartphones or the central operating systems, like Windows, Apple's Mountain Lion, Linux, and its variants—is revised soon after its release, with updated versions automatically sent to users. Although the companies extensively test their software before launching it, only the experiences of millions of people, under all combinations of circumstances, are adequate to reveal significant problems. That point is taken further under a famous maxim of open-source development, called Linus's Law in honor of Linus Torvalds, the mastermind of Linux.

It holds: "Given enough cyeballs, all bugs are shallow."

That is, crowdsourcing would not only identify the problem but help fix it. That was often more article of faith than established fact; the vast majority of open-source projects had only one or two people working on them, even though their code was publicly available. But there were a few historic exceptions, most prominently Linux and Firefox. As it happened, the volume of contributors to Diaspora was extraordinary. And they had plenty of things to fix.

The plan was to send out the first batch of one thousand invitations on the Tuesday before Thanksgiving, giving them Wednesday to get out of town and on their various ways home. Each of those invited would have five invitations to distribute. Most recipients were not software developers but civilians inspired by their cause. Real people.

He wrote: "We are proud of where Diaspora is right now. In less than five months, we've gone from nothing to a great starting point from which the community can keep working. We've spent a lot of time thinking about how people can share in a private way, and still do all the things people love to do on social networks."

Then he broke news about a feature of Diaspora that would distinguish it from existing social networks: "Diaspora lets you create 'aspects,' which are personal lists that let you group people according to the roles they play in your life. We think that aspects are a simple, straightforward, lightweight way to make it really clear who is receiving your posts and who you are receiving posts from. It isn't perfect, but the best way to improve is to get it into your hands and listen closely to your response."

This was a striking, if obvious, innovation. Up to that point, anything posted on Facebook, be it a picture or a comment on a news story, was visible to all of a user's "friends." It was possible to block the general public from seeing it. But Facebook and most social networking sites did not allow users to exercise anything like the kinds of granular control real people used in their everyday interactions—the instantaneous, natural grading of communication with different individuals, the changes in inflection, in tone and topic, when speaking with members of a softball team or friends from work, high school classmates or close relatives. Diaspora introduced a simple way to allow the world to be divided beyond the crude boundary of "friends" and "nonfriends." As people were brought into the network, the user could tag them, essentially sorting them into different buckets. A post could be distributed to any or all of those buckets. That meant a message meant for drinking buddies would not end up in the news feed sent to Mom and Dad. The subgroups in the buckets were called aspects, as in the different aspects of one's life.

The creation of aspects was a necessary contribution, and much in keeping with their goal of not just improving geeky stuff like privacy and data ownership but also making communication richer and deeper.

The post, signed by all four, promised that they had fixed the security bugs that had gotten so much attention in September, and that they were going to continue to tighten the hatches. They also were going to put together better documentation, clean up the code, and make it easier for all the people hosting Diaspora pods to upgrade each improvement. The post concluded: "Our work is nowhere close to done. To us, that is the best part. There are always more things to improve, more tricks to learn, and more awesome features to add. See you on Diaspora."

The tech blogs exploded with the news, and tweets began flying. The announcement had been a bit premature: before they could distribute it, they needed signed agreements from the people who had contributed code, agreeing to release their copyrights to Diaspora Inc. This took e-mail hand-holding because the license they were using was thought to be a mess. Finally, though, the "pre-alpha" version was released late that evening.

The response was vigorous and lively, the hissing and sneers drowned out. They had kept their promises: to release a version to developers in September, and now, two months later, an alpha for the public. Diaspora was not vaporware. It took a little while for the public to absorb the reality that Diaspora was not a website; it was a piece of software that ran a website. Someone had to host the website, provide the servers, maintain them. The supporters who received the early invitations could sign up on a website that the guys themselves were hosting, JoinDiaspora.com. But the guys had no intention or desire for this to become the only place where Diaspora existed. Anyone could host a site. The software was free. While it was raw and buggy, precisely what would have been expected at that stage, work on it would continue. All fixes would be shared with anyone hosting an instance of Diaspora. The very nature of their plan was that they would not be in charge of everyone's data: they were drafting the DNA of a distributed social network, not growing and feeding the creature.

The evening after the release, the four guys began to scatter to spend the Thanksgiving holiday with their families: Max was going to relatives in Seattle, Rafi was heading to Palo Alto, Dan back to Long Island, and Ilya was pointed toward the suburbs of Philadelphia. Before he left, Ilya grabbed Max.

In the weeks since the visit to New York, Ilya had lost just about all interest in the monastic regimen in San Francisco. Even though he had just found the apartment in the Mission District with an amiable house-mate, the grind of long hours, the absence of friends outside work, and the months in the depressing hotel had spent all his emotional reserves. Ilya seemed to assimilate the decor of the flophouse. He shaved only occasionally, about as often as he showered. He wanted out, he wanted to return to being a college student in New York, and he told that to Max.

"I'm going to wait it out until Thanksgiving," Ilya had said.

That, in effect, was the following day, as they would all be going home. Max was furious.

"What are you going to do? You can't just leave," Max said. "You have to give me until Christmas to get someone to replace you."

Ilya could not tolerate a horizon that long. He wanted to restart his life in New York. The man-boy who had been so exuberant, so loud, nine months earlier when he and Max were talking on the phone about their plans to drop everything and make Diaspora that the neighbors had to yell at him to shut up had, it seemed, run out of gas. He was flat.

It was a particularly bad moment for this crisis. Managing the demands of the users who had just gotten the invitations, fixing the bugs that they would spot, would take every BTU of brainpower among the four of them.

"If you leave tomorrow, I will be really mad. But if you give me two weeks' notice, we will get someone to replace you," Max said. He retreated on his demand for Ilya to stay until Christmas. "If you tell me you want to go, give me until December 1."

Ilya mumbled something, seemed to agree. Max put the thought out of his head.

Through the night, they watched the site come to life, people joining every minute. Some mentioned they wanted private-messaging ability, instead of having to put everything on the walls of their aspects. A developer in Paraguay checked in and said he would work on building such a feature over the weekend. Serbian users said that they were going to translate the code immediately. A collective in Spain was organizing a meeting. Max and Dan were up all night, replying to comments, pointing people to places where they could file bug reports.

Back in New York, their mentor, Evan Korth, posted the news on his Facebook page: "Diaspora FTW"—in online shorthand, Diaspora For The Win.

In the morning, Dan headed back to New York. At the airport in San Francisco, he went through a new screening device run by the Transportation Security Administration. He sent a quick note about the experience to his Diaspora page, in the grand tradition of social network ephemera: "TSA just saw my penis."

Just as in America the fourth Thursday in November is a ritual of turkey and gratitude, in Egypt that weekend would mark a kind of ritual: the parliamentary elections, which had faithfully returned the party of Hosni Mubarak to power, as it had in every poll since the party was created in 1977. The activists whose agitation had grown louder online and had, on occasion, moved into the streets were not mounting a serious slate in opposition. The rigging of the election was not something they worried about; it was part of the routine. Participation under such circumstances was futile. They decided, instead, that they would document every instance of fraud on a prized, uncensored forum: their Facebook page.

Just a few days before the election, they came up with a bold plan to at least thwart some of the fraud in which ballots were cast in the names of nonvoters. Everyone who was registered would cast a write-in ballot for Khaled Said, a young man killed during an encounter with police.

On that Friday morning, Wael Ghonim awoke to discover that Facebook had unilaterally and without notice taken down the "We Are All Khaled Said" page. A second page, "Mohamed ElBaradei" also was shut down. The news burst on Twitter. It reached across the ocean. Egyptian activists in the United States and Europe tried to contact managers of Facebook, an institution that on regular workdays had no public address, e-mail address, or phone number. On the Friday of a holiday weekend, the hurdle was even higher, because even personal contacts were likely to be away. Nevertheless, by the end of the day, they had lined up people who were in a position to act.

Their reasons for shutting down the pages were not, as many in Egypt had speculated, the result of pressure from authorities. Facebook had received complaints about abuses on the pages, which was not unusual. But the administrators of the pages had used false identities, a violation of Facebook's terms of service. Knowing real identities of their users, and their appetites, was Facebook's business. The company also held that real identities provided accountability for what was said and done on the site. The activists, of course, believed that whatever the merits of Facebook's position, "accountability" in the hands of the Egyptian regime would mean suppression and retaliation.

"Our attempts to remain safe," Ghonim wrote, "had come back to handcuff us."

Scrambling to get their pages back up before the elections, they made a deal to have activists outside the country be the administrative contacts. Facebook promised that it would not reveal their names. The "Khaled Said" page and its 330,000 registered users, and the "ElBaradei" page with 298,000 users, were back within fifteen hours. It was Facebook's world, and they set the rules.

That weekend, as she listened to the cell phone rings roll on and on, Stephanie Lewkiewicz wondered if he would pick up.

At last, she heard a human speaking.

"Hey," Ilya said, barely getting the word out of his mouth.

She had dialed his number and that was his voice, but could these flat, lifeless tones possibly be emerging from the same guy Stephanie had seen two years earlier at the front of the math classroom, practically floating from his seat with excitement?

Was he okay?

"I've quit Diaspora," he said.

Ah. That explained why he was down, but it did not add up to world's end. Diaspora, hacking, building, activism—whatever he took on, she had absolute faith in him.

Two years earlier, Stephanie had realized that she no longer enjoyed being a theater major and that even if she slogged through another couple of terms to graduation, she would join armies of people struggling to keep themselves together while waiting for a break in the brutally competitive theater world. She didn't love theater that much. In fact, she didn't love it at all. She was a natural at math—her father was an engineer—and saw her relationship with that subject as love-hate. Which, at least, was better than theater.

For her senior year, she decided to double major—bring on the math—and shifted into the prestigious Courant Institute at NYU. Many of the Courant professors were too absorbed by research to do much hand-holding; it seemed the students were essentially on their own. Most of her Courant classmates were more introverted and socially awkward than the theater crowd, but she felt they were also more innocent.

Studying kept her up all hours. One night, after crashing on a home-work assignment in the math lounge, she curled up on a sofa and closed her eyes. At some point—she didn't know how long she had been asleep—she heard a voice.

"Hey there," a guy said.

She blinked her eyes open and lifted her head.

"Oh," the guy said. "I've seen you around. You're in my analysis class."

Stephanie recognized him right away. He was the guy who wore a crazy American flag shirt, tie-dyed pants, and sat in the front with his hair sticking up at all angles. How he looked was the least of what made him memorable. Generally, few people in math class spoke up, and of those who did, Stephanie sometimes felt there was a touch of showing off how much they knew. This guy was always engaging the teacher, but it was different. He was openly, promiscuously ravenous to learn, taking delight in everything about the subject. An explanation, a link made clear, would make him laugh. She had found him compelling.

"Oh, yeah," Stephanie said. "I've seen you in class."

"I'm Ilya," he said.

Good thing he had rousted her. Analysis, a course in advanced calcu-lus, was about to start. They walked to class together. The conversation was a bit forced, but by the time they got there, they were shooting beams at each other. In the room, where it looked as though the other thirty students were staring at their desktops, they split up and took their usual seats—Ilya in the front, Stephanie to the side and back a row or two.

Throughout the class, he turned to her and smiled.

After that, they were together all the time. In math club. At special talks or any other gathering. All-nighters in the math lounge. He wanted to meet everyone, find out what people were up to. On weekends, he might call and ask, "Do you want to get lunch?" It was best not to arrive too hungry. She'd get to his apartment and he would be up to his elbows in a box of electronics.

"Can you just hold this for a minute?" he'd ask, and they would forget food while she gripped pliers and he wielded a soldering iron, tinkering until he'd had his fill. The tools of mischief, rather than the making of it, were compelling to him, and there was always something to make. For instance, once he'd heard about TV-B-Gone, a remote control that could switch off any television—in, say, a waiting room, a bar, a student lounge— he had to have one. It could be bought either ready to go or as an open-source kit that had to be assembled. Of course he bought the kit and put it together. Though he never really used it that much.

In groups, he'd spiel about limits to freedom on the Internet, and any other social causes that had caught his interest. He liked to hang out at Bluestockings bookstore on the Lower East Side, a café, center for activists, and seller of books that looked at the world through a radical lens, or nearby at ABC No Rio, the collective run by artists and activists.

Stephanie usually did not accompany him on these journeys; they had no exclusive claim on each other. Having grown up as the immigrant outsider in high school who had to wipe the sweat from his palms so he could shake hands with one new person every day, he seemed to relish the life of a nerd Lothario. He loved that girls found him and his zany ebullience to be winning, and that they seemed to remain fond of him even after a dalliance of short duration. "A single-serving friend," Ilya called them, invoking the movie *Fight Club,* where the Narrator uses those words to praise fleeting but valuable conversations with people he meets on airline trips.

His relationship with Stephanie was deeper, richer, plainly romantic, but it was also beyond any category she could name, with neither formal nor informal boundaries drawn. In the summer after they first smiled at each other, it turned out that they both had commitments that kept them at school in June and July, so, by default, they continued to spend lots of time together rather than drifting away. But they never met each other's families.

As Ilya migrated from the math department lounge to hanging out in the computer club room, and working on the MakerBot, Stephanie got to know his new friends there. He was always peeling off to make some repair on the 3-D printer, and even gave her a lumpy thing that was an early product, though she never quite figured out what it was meant to be. Whatever he had going on, she wanted to know about it. She loved hearing him rail about a social issue. So many people were just thinking about getting good grades and passing classes and partying. He had his eye on the wider world, and was passionate about it.

When the Kickstarter campaign went on to its roaring success, she felt that the world was responding affirmatively to a person who loved it, and who would make good use of the support.

That episode also provided a kind of punctuation mark to their own, not-quite-recognizable relationship. She was graduating, a bit late because of her double major, and pointing herself toward graduate school. He was two years younger, and was heading off to San Francisco to spend the summer, and perhaps longer, working on Diaspora. To make parting simpler, they decided that they would not speak until the fall. This was his initiative. They'd had a few calls in September; he was buzzing over the release of Diaspora. They managed to squabble a bit, but it was nothing serious.

Now, though, nearly three months later, on Thanksgiving weekend, he barely answered questions about how he was. For a moment, Stephanie thought that it might be the tiff they'd had a few months earlier. But that had been nothing—she'd almost forgotten it. His quitting Diaspora was a surprise, but he had so many options.

"So what are you going to do?" she asked.

He was going back to NYU with a big proviso.

"If they'll take me back," he said.

"Why wouldn't they?" she asked.

In his funk, Ilya had assembled a psychic railroad train that ran on unyielding tracks of despair, each boxcar filled with some new worry or anxiety. Had he followed proper procedures in setting up his leave of absence? Maybe not. Then he had just extended the leave, but there wasn't a formal approval process. So maybe they wouldn't regard him as a student anymore, just as some guy who had abandoned the program, a dropout. Maybe they would make him reapply. He might not get in. And so on.

Stephanie thought the particulars were, one by one, ridiculous, but she also recognized that the logic of depression was impregnable. She had been close to people with serious psychological problems. Yet in the two years they had known each other, she had seen Ilya only with the lights on, fully ablaze. Perhaps she could talk him out of the shadows. She had never gone to his family home in Lower Merion Township, just outside Philadelphia.

"How about I come down to Philadelphia to see you?" she suggested.

No, no. He would be fine. He just had to work out some details of re-admission to school. They signed off. Stephanie was profoundly unsettled.

Ilya's gloom was disconnected from the reality of what the four of them had pulled off since the night of Eben Moglen's speech back in February. Diaspora worked.

During the first hours and days when it could be used and looked at, there was no rush of apologies from the prophets of cynicism who either had accused them of running a colossal scam or had no hope of their building something people would use.

Sign-ups moved at a brisk pace during Diaspora's first semipublic weekend, even though the Diaspora Four had taken pains to titrate the growth, mindful of Friendster's collapse a generation earlier under the weight of its own popularity. Invitations went out first to Kickstarter contributors, each of whom would be able to issue five invitations. Once that group had reached its mass, the team would begin issuing invites to more than 200,000 people on a waiting list that was growing by the minute.

Facebook had also inched into the world at birth in 2004. At first, only students with a Harvard.edu e-mail address could sign up. Then it moved campus by campus across the country. Six weeks after the launch, it had twenty thousand members. Not only did the slow rollout give Facebook a discrete, technically manageable load, but each school provided a ready-made, built-in social network where the users knew one another or wanted to.

Diaspora, by contrast, did not have such organic networks. Its users were bound by their support of the idea of it, their ardor for privacy, their love of free software, or their abhorrence of Facebook. The early reviews of Diaspora praised the simplicity of its interface, or at least were neutral about it, but were bewildered by the absence of actual friends or any way to find them.

"What if you build a social network and no one is there?" Christina Warren wrote on *Mashable.* "Here's the big problem with Diaspora—no one is there. . . . Right now, it's like adult swim at the YMCA. You know there aren't any adults at the pool so there are just a bunch of kids standing around the empty water."

She added: "What makes Diaspora worse is that there is no easy way to find people."

On *ZDNet,* another important tech site, an article was titled "Is Diaspora Too Late?" The author, Dana Blankenhorn, was skeptical that the project could move ahead with just its current foursome.

"The challenge for Diaspora—for any challenger—is convincing masses of people to try a second social network," Blankenhorn wrote. "If people can be convinced to join, then Diaspora has to scale its development process. Facebook is organized. It has gone through that process. You're not going to maintain a competitor with just the four partners, no matter how well they code."

In the comments section, one contributor said it was missing the point to gauge Diaspora's success on whether it became a business. "No one cares about Diaspora—the website. It's all about the software. It doesn't matter the least bit if they manage to make a buck off it—the software will always be continuously developed by its users—be it single users, or large companies trying to monetize it. Just look at the amount of code contribution they already claim."

And on *Ars Technica,* Ryan Paul noted the rawness of the first release, but said it held out a great deal of promise—though not necessarily as a

Facebook killer. "It may seem far-fetched, but Diaspora (or something like it) could someday help to inspire change in the social network arena in much the same way that Firefox has helped to reinvigorate the browser market and accelerate conformance with open Web standards," he wrote.

"Diaspora doesn't have to topple the entrenched giants in order to inspire positive changes in the industry; it just has to get a critical mass of people to start thinking more seriously about privacy issues and the right kind of interoperability."

On the Monday after Thanksgiving, Max returned to the Pivotal offices with a long list of things to do. The release of the alpha version meant that they had kept the second of the two large promises: to put out a working set of code for other developers to hack on, and to run their own node for supporters.

The Italian edition of the fashion magazine *Vogue* was calling to set up a photo shoot; Daniel Grippi was of unmistakable Italian descent. A company that tracked the number of people working on open-source projects found that Diaspora was consistently among the top two or three in the world. Every day, on average, nine hundred links on the web pointed to Diaspora. The next largest open-source project on the web had three hundred, and most had around thirty to forty-five. "It's quite astonishing," said Philip Marshall, an executive with Ohloh, which ran a search engine that tracked open-source projects. They were declared rookies of the year by Ohloh, and one of the top ten start-ups by *ReadWriteWeb*.

And they even inspired a small cultural war when a person who went by the name of Avery Morrow, an early supporter of the project who had also been documenting their progress in a blog, publicly renounced it. He took exception to their failure to just have a simple drop-down box for users to choose "male" or "female" in their profiles. Instead, they had left the box blank so that people could identify themselves. Wrote Morrow: "This is a sign that the programming team—not some unrelated pinheads, but the five or six people who are supposed to be writing the code—have put strong, usable code last on their priorities."

Sarah Mei, the developer with Pivotal, had created the gender box and explained why on a web post.

"I made this change to Diaspora so that I won't alienate anyone I love before they finish signing up.

"I made this change because gender is a beautiful and multifaceted thing that can't be contained by a list. I know a lot of people aren't there with me yet. So I also made this change to give them one momentary chance to consider other possibilities.

"I made it to start a conversation.

"I made it because I can.

"And, of course, I made it so you can be a smartass."

Writing on the blog *Econsultancy,* Patricio Robles said it was a sign of how unserious Diaspora was, compared with Facebook.

"Facebook develops new features, of course, based on an analysis of real-world usage, and when it ships new code, it iterates as necessary based on the feedback it receives from real users. Diaspora, on the other hand, isn't even out of private alpha and one developer has already *single-handedly* decided how one of the most important fields will function with implications for data consistency, search and usability being brushed aside. 'To start a conversation,' and because she can.

"That's not likely to be the foundation of a successful consumer internet product."

Perhaps the most striking reaction to their arrival was an article in *Scientific American* published just the day before their alpha release, in an issue marking the twentieth anniversary of the invention of the World Wide Web. Its principal creator, Sir Tim Berners-Lee, bemoaned the increasing strictures that were shrinking the autonomy and the privacy of individuals. He saw the silos of giant social networks draining the web of its vitality. He hailed Diaspora as one of the projects making a path to a healthier web. Berners-Lee had transformed the world. He had never attempted to patent his conception. So this pack of pizza-eating, sneaker-wearing, scruffy college kids was being praised for their vision by an authentic giant.

None of them even knew about the Berners-Lee salute; they were absorbed with other, more immediate developments.

After Dan got off his flight, he called Max from the San Francisco airport to let him know that he was on his way downtown.

"What's going on?" he asked.

"I'm here working with Rafi," Max replied.

"What about Ilya?" Dan asked.

"I have some news," Max said.

Ilya had already called that morning.

He wanted to go back to school, and he would be checking with NYU to see how he could resume classes in the new year. So he wasn't returning to San Francisco. Max replayed the conversation that he and Ilya had had the week before, when he thought he had extracted a promise from Ilya to stay with the project until at least the beginning of December. He was stunned by the call.

"I was like, 'Fuck you, dude,'" Max said.

They were getting crushed by work. Ilya might not have been a great coder, but they had no end of tasks that would not tax his skills. Even just keeping track of the submissions for revised code, or getting feedback from the Real People now using the site. It was treacherous for Ilya to walk with no notice. Striding through the gleaming arrival hall of the airport, Dan screamed obscenities into the phone.

# CHAPTER FOURTEEN

By the late fall of 2010, an undeclared, all-out worldwide war was raging across cyberspace. A private in the U.S. Army had downloaded gigabytes of classified intelligence communications and passed them along to the operators of a website called WikiLeaks. Thousands of classified cables and documents written by U.S. diplomatic and intelligence officials were posted by WikiLeaks.

Almost immediately, a massive cyberattack began on the Swedish provider that hosted the WikiLeaks website. The attackers were not identified, but the website was bombarded with requests for communication at the rate of ten gigabytes per second. The requests were generated by robots, not people. The provider of the hosting service had to shut down the WikiLeaks site in order to preserve the rest of its business. Peter King, a New York congressman, urged the Obama administration to declare that the site and its best-known figure, Julian Assange, were terrorists and spies. WikiLeaks then moved to Amazon's cloud service because its structure had vast elasticity and the robustness to absorb the cyberattacks. Senator Joseph Lieberman said that he planned to question Amazon about its relationship with WikiLeaks. Within a day, Amazon canceled its hosting relationship with the website. WikiLeaks went back to its Swedish host, but then lost its DNS provider, the system that translates a domain name into a physical address. It then moved to a domain controlled by the Swiss Pirate Party, one of a hub of digital libertarian, anticopyright political organizations on the rise in Europe.

"Despite its name, 'cyberspace' runs on physical infrastructure that sits in various governmental jurisdictions, and when sites like Wikileaks start irritating those governments, sovereignty is quite power-fully brought to bear," Nate Anderson wrote on the technology website *Ars Technica*.

A collective of digital activists—hacktivists—operating under the flag of "Anonymous" began what they called retaliatory strikes against the websites of governments and corporations that were at odds with WikiLeaks.

On December 3, the three remaining Diaspora partners opened their e-mail to find a message from Ilya, his first communication since he'd called Max to say that he wasn't coming back.

The subject line was "Thank you for understanding."

He had been to two doctors, he reported, and although there was no final diagnosis, it appeared that he had developed over the previous three months some unspecified "medical conditions" and the doctors had ordered him not to make any "life-changing decision." He presumed the doctors would fix his problem within the next four weeks, and he would be getting plenty of exercise. They should hold back on his December allowance and use it for the project in the future.

In accordance with his doctors' recommendation, he wrote, he wanted to withdraw his resignation, as "it might be erroneous due to my medical condition." After four weeks, he said, he hoped to be totally fine.

"I need your support now," he concluded.

There was nothing for the other three to do but go back to work.

About a week later, Stephanie, worried that Ilya was locking himself away from the world, prodded him in a surprising way: she posted a link on his Facebook page for a party and lecture marking the seventy-fifth anniversary of the Courant Institute, which housed the molten core of mathematics at NYU. Moreover, the featured speaker was Michael Shelley, a professor of math and neural sciences for whom they both had huge regard.

"Come!!" she wrote. "Mike Shelley, who could say no? :D"

He did not reply to her entreaty, and when she phoned him a few days later, he declined again.

As she had before, she suggested that she would take a road trip instead to Philadelphia.

"Oh, no," he said. "I'm fine."

Back in San Francisco, in the apartment Ilya had just moved into, his roommate, Gardner Bickford, had not seen him in weeks. Gardner texted and asked Ilya if he wanted him to find a new roommate. Despite his absence, Ilya had been prompt with the rent. He told Gardner to hang on, he'd be back in January.

His partners at Diaspora were not so sure.

For Christmas, Dan bought himself a better pair of headphones than the ones he already had. At work, they blocked out the world, kept him thinking about only what was in front of him on the screen and no peripheral conversation. "I don't know what I would spend money on," he said over a beer. "I don't have time to spend money. All I do is work, think about the project, and sleep."

Dan Grippi would never strike anyone as monk material, but he had left himself behind working on Diaspora, burrowing into a kind of hermitage. In high school, he had run cross-country, a test of solitary endurance; at Diaspora, he spent much of the day in front of his screen, speaking little, a reticence that could be mistaken for arrogance instead of shyness.

During the break, his father had been speaking to him about taking his skills into the business world. Banks were always looking to hire people like him. Dan was not interested. "I don't see why someone would go to college just to get out and have someone else tell you what to do," Dan said. "And we have had it easy—we actually got everything handed to us on a silver platter."

Now the project was gunning along. New fixes, better features, a trickle of new invites distributed every few weeks.

The departure of Ilya had infuriated him, in part because he, too, was so unhappy in San Francisco, and they had relied on each other in ways that the others did not have to—the Sofaers were in northern California, and Max's high school buddy Dan Goldenberg had moved in with him. By happy chance, other friends of Max were also in the Bay Area. In walking out, Ilya had not only left the project one man short but ditched Dan on an emotional level.

"I took it personally," Dan said. "Rafi has his brothers. He has his family in San Francisco. So great, he's back home. Max has all of his friends in San Francisco. I know absolutely nobody. I'm busting my ass. Same exact thing, similar situation. I'm dealing with it just fine. Then he just bails? No way. Totally fucked. It is what it is."

There were negotiations afoot to allow Ilya to come back on a provisional basis, to see if he could tolerate it for a week or so. And at the request of Ilya, Mike Sofaer also was urging the three guys to let him back. Dan was still opposed. If the guy was sick, he needed to be home. The Diaspora project was not a way to convalesce.

Dan could not stand living in San Francisco, yet when he was back east for Christmas, he counted the minutes until he returned. He had been unable to detach himself from the all-absorbing demands of Diaspora. Why not do it remotely and collaborate via computer with Max and whoever was in San Francisco, and have a life in New York?

"As much as I love New York, the project needs me in San Francisco," Dan said. "You're going to be left out of the loop if you're not in the same spot."

All hands were back in the Pivotal offices at the beginning of January, Ilya included. It was more than passing awkward. Seated ten feet away, Dan did not speak to him for a solid week. The freeze-out ended when Dan called a meeting at a coffee shop to discuss the future of the project. They ended up on the same side of an argument. By Monday, they were pair programming together. Ilya was reengaged, but not single-mindedly. Since discovering Kickstarter for Diaspora, he had pitched in small contributions to other campaigns, including one for a rooftop farm in Brooklyn, a project to record classical musical and release it without copyright restrictions, and a book on punk mathematics that promised to explain "better living through probability" and "using orders of magnitude to detect bullshit."

Now something else had caught his eye: Eben Moglen had turned to Kickstarter to raise money for his Freedom Box project, the pocket-sized server that people would be able to plug into a bedroom and use for all their communications. It was the ultimate in federation or decentralization.

Ilya posted a note on the Diaspora page: "With much nerd love, just backed The Freedom Box."

There was not much direct conversation between Ilya and the others about his departure. Mike Sofaer had worked hard to talk Dan and Max past their fury. Yes, they were right to feel jerked around, and that they had been left hanging.

"Look," Mike said. "It isn't a huge risk that it will happen again."

Soon, Ilya was absorbed again by Diaspora—and also by roiling developments on the other side of the world, introduced to him by Yosem Companys, the group's cheerleader and friend.

In the first months of 2011, as the uprising in Egypt accelerated, Yosem found himself working all hours to run the Twitter account for a group called Liberation Technology, which consisted of people interested in deploying emerging technologies to bring about social change. Their virtual commune was an e-mail list and Twitter. When Hosni Mubarak's government shut down the Internet ahead of a planned "Friday of Anger" on January 28 in Cairo's Tahrir Square, the Twitter account for Liberation Technology almost instantly became part of a digital bypass.

Activists in Egypt—and other countries, like Libya, Yemen, and Bahrain, where revolution was in the air—were still able to get messages through to Twitter (in fact, Google and Twitter set up a relay system, in which voice messages left on a phone system operated by Google were transcribed and posted to Twitter).

Yosem would pick up urgent messages and relay them to sympathetic hackers, including, most prominently, Jacob Appelbaum, a leading security researcher. There were many others, including Ilya, who fielded requests for technical help. Sometimes Ilya helped keep servers going when they came under attack. He also helped the activists regain Internet connectivity via satellites. He wound up conferring with an Egyptian technologist he'd met at a conference in Berlin.

He set up an Arab Spring Diaspora pod. "Some activists used the dedicated Diaspora Arab Spring pod to communicate because it was not as prominent as Facebook and Twitter and thus could sneak under the regimes' radars when the regimes started cracking down on Twitter and Facebook," Yosem said. But there was no technical stealth built into the Diaspora code, he noted: "It had everything to do with governments' ignorance of Diaspora. In fact, Diaspora was riddled with holes and was not a secure solution. It was just a good enough decentralized one." Not long after the Arab Spring roared to life, three Diaspora pods turned up in China.

"Did you see these?" Yosem asked Ilya.

"Yeah," Ilya said. "I met some Chinese activists online, and they're using it to circumvent censorship."

Always cautious about his own position, and jeopardizing the status of Diaspora, Ilya shrouded this part of his life. When he and I next met and talked about the events in the Middle East, he mentioned that he was keeping track of what was going on, but did not volunteer that he had been providing assistance, however limited, to the activists. Max, Dan, and Rafi knew nothing about it. In fact, a month after Mubarak shut down the Internet, Dan and Rafi were back in New York to speak at a free-culture conference. During the question period, a man asked: "Have you ever proposed that the Diaspora tool could be a way to communicate during moments of protest? A space that is safe to share information, sort of off the radar?"

"Diaspora right now is not a system made to be operated in an environment where people are trying to take it down," Rafi said. "Hopefully, one day it will be able to operate at that level of robustness."

Plus, he said: "What makes a system useful in those situations is really social penetration. The fact that a lot of people are on Facebook, that a lot of people are able to see what is going on on Twitter, makes them useful in those hazardous moments.

"I don't think that's our target market right now—revolutions."

Well, someone else asked, how about setting up a mesh network— a kind of local Internet, rooftop to rooftop, not vulnerable to centralized shutdowns like what happened in Egypt? Rafi politely sidestepped the role of digital messiah.

"I think that's a great idea; it's something I'd like to work on someday.

It's not what I am working on right now. I think that New York City is a good place to experiment on with that, because of its density."

He paused for a second, then lobbed the suggestion back to the audience.

"And—do it."

Toward the end of January, the four guys were given an invitation to an afternoon rubbing shoulders with wealthy young investors and hackers who hoped to get that way in a Silicon Valley incubator called Sunfire Offices. It had been started by former engineer managers for Google and Facebook. On Thursdays, they held mixers with promising start-ups, including people working in the Sunfire space and young people floating around with small fortunes made in technology who were looking for new ways to engage with the world. Invitations were hard to come by, and the Diaspora crew was thrilled to be asked. A company called Flipboard would be giving a presentation before the social hour.

"You should come with us," Max said to me.

I volunteered to pick them up at the Pivotal offices and drive to Mountain View in the valley. A few minutes after two, parked on Market Street, I saw Max walking across the street by himself to the car. Max opened the door. He was anxious.

"I'm really sorry about this. I got ahead of myself," he said.

Across the street, Dan, Rafi, and Ilya stood peering at the car.

"Why are the other guys standing over there?" I asked.

"I didn't exactly clear it with them before I asked you to come along," he said. "They're really mad at me. I'm sorry. I'll pay you back for the rental car. They don't think it would be cool for them to roll up with the *New York Times*."

He apologized repeatedly.

"I've got the car anyway," I said. "I'll drive you down. We can chat along the way."

He crossed the street to deliver the invitation. All four piled into the car. Hardly a word was uttered on the forty-five-minute drive.

# PART TWO
## Yo, This Is Our Internet

# CHAPTER FIFTEEN

A round a sleek conference table, jotting notes on legal pads and nodding, four partners from the venture capital firm listened to the pitch. The four Diaspora dudes had worn blazers and ties with their jeans for the occasion, which initially prompted a bit of ribbing from the ever-casual capitalists. Among them was one of their earliest and most persistent boosters, Randy Komisar. Not only was he not bothered by their lack of polish, it was part of what made them endearing. It was their first pitch meeting, and they were in the offices of Kleiner Perkins Caufield Byers, the JPMorgan Chase of Silicon Valley. Diaspora wasn't passing the hat for ramen money anymore. After the teasing had eased, they got down to business, running a slide show about the project and what they believed it needed to keep going. It was not until the seventeenth slide that they reached "the ask"—the amount of investment they were seeking from venture capitalists.

Kleiner Perkins was poised to hurl money at them. A few weeks earlier, Komisar had e-mailed Max, telling him, "Get in here," Max reported. They made a date for the afternoon of March 8. They were turning a corner. The project—fueled by pizza, the buzz of being awake at two A.M., the giddy plans to eat ramen and code for an entire summer in Max's family's basement, the thrill of building something disruptive—was over. Everything promised to their Kickstarter backers had been delivered.

For weeks, Max, who had assumed responsibility for getting their pitch together, just as he had organized their Kickstarter campaign, struggled to figure out how much funding they should seek. He immersed himself in the vocabulary of start-up speak, the meaning of the various levels of funding, what a seed round meant as opposed to a Series A, and how each dollar of financing would shave another layer from the stakes of the founders. Given what they had pulled off with a little money and four guys who had no significant experience, Max believed, they had shown that Diaspora was for real. But he also felt sure the four guys had gone about as far as they could on their own, even with open-source contributions coming in at a robust rate. They needed to hire experienced engineers who could improve the code and make it run smoothly behind the scenes, and also bring sparkle to the features that the users saw. That took money.

He was sure they were in a ripe moment. The early, exuberant responses to Diaspora—even the idea of it—were like the flood of affection for Firefox when it emerged in 2004. He had a gift for reading history and seeing its reincarnational possibilities. From the folkloric history of Netscape, which the marketplace viewed as a colossal financial failure, the gaseous product of a twentieth-century tulipmania, he drew a more enduring lesson, of how a mythic, Internet-changing company liberated the web from the control of Microsoft, even though it was destroyed in the process. With huge venture capital investments, it had hired every available software engineer and designed iterations of iterations. Netscape did not just throw money at problems; it bombarded them.

"I don't necessarily know if we're Netscape yet, but we could be that place," Max had mused a week before the pitches. "It has just captured my imagination recently. How do we promote sharing in a way that doesn't totally screw you and makes the Internet a better place? How do I build a place where I can bring these insanely smart people to come and love Diaspora? That's what it's all about. Finding people who are as passionate as we are, and building a whole organization of people who want to help us build the next version of the social web. We need to get this little hump of pitches out of the way." During the planning, he called Yosem. For Max's purposes, Yosem had the highly practical credentials of working in consumer products at Procter & Gamble, and building online political organizing tools during the 2004 presidential campaign. Plus he coached people on how to make pitches.

Yosem had worked as the teaching assistant at Stanford with Randy Komisar, in a course that taught entrepreneurs the nuts and bolts of making pitches. They spoke about basic elements, like identifying a need or hole in the market, and how Diaspora would solve it.

"What do you think we should ask for?" Max asked.

"You should leave it open-ended," Yosem said.

Max disagreed, saying that the group believed that an ask price was needed for the pitch, or else they risked looking like they did not know what they were doing. That did not make sense to Yosem. After all, they didn't know what they were doing.

"You can't pretend if everyone, especially investors, knows you have no prior experience in that area," Yosem said.

Even so, Max insisted, they couldn't present themselves as naïve kids with some kind of summer project. So Yosem called a friend at another venture capital firm, and ultimately decided that the capital market for consumer web applications would support an initial investment between $2 million and $3 million.

The night before the pitch, they stayed at the Sofaer home in Palo Alto. The group was not happy with Max's first draft, and set out to revise the deck of slides. The opening sentence took six hours: "Diaspora is an open-source social-networking ecosystem that gives users control of their data and social graph."

It wasn't poetry, but having it out of the way allowed them to move on to the rest of their presentation—including statistics about dissatisfaction with Facebook's privacy settings, and the traction that Diaspora had gained among developers, 4,000 of them following the project on GitHub, more than 100 of whom had contributed code changes. It had been translated into 32 languages. There were 350,000 e-mail addresses waiting for invites, a number that was growing. Some 48,000 people had liked them on Facebook, and 65,000 were following the project on Twitter. Just on their own pod, JoinDiaspora, they had 43,000 registered accounts, and of those, about 5,000 were active—meaning that they used it at least once a week.

"All we did was write a few blog posts (no marketing whatsoever)," one slide read.

"We work fast, have people's attention, and an exploding community—all on less than $200K."

Yosem stayed with them for much of the night. With some dismay, he realized the presentation was a mess and had to be redone from scratch. It was like a homework assignment at NYU that was due the next morning with a $2 million grade potential. Still, he brought a center of gravity to the proceedings, his presence helping to dampen intramural squabbles over single words. He had volunteered to accompany them to the pitch meeting and help translate the VC-speak, but Max had said that only the original team members would be going.

There was one other area where Yosem's advice was sought but not heeded. On the slide for the ask, Max had written $10 million.

Yosem was flabbergasted. It was outlandish. His own counsel had been not to put any number in, but this number was five times what investors had suggested would be feasible.

"How did you come up with ten million dollars?" Yosem asked.

Max said that in preparing for the pitch, he had been consulting with Mitch Kapor. The figure made no sense to Yosem.

"It seems outlandish, especially since the team does not have any business experience," Yosem said.

As the evening grew later, and the pitch began to take final shape, Yosem tried one more time to get them to be vague about the amount of money they were looking for.

"Take out the ask," he said. "Leave it open-ended."

"If we don't have it in, we're going to look very inexperienced," Max said.

Moreover, one of the start-up gurus, Paul Graham, the subject of Max's never-completed senior thesis, had written that fledgling businesses should make their pitches assuming that they would never be able to raise money again. "It's the opportunity of now," Max said later. "People are already screaming bubble."

He'd thought about the coverage of their funding if their pitch succeeded, and the outrage over the extravagance of paying Diaspora. Soon, he was sure, there would be articles about the tech bubble, and how investing in Diaspora was an example of the lack of discipline.

"We will be one of the five points that show we are in a bubble," Max said.

They all were up early on the day of the Kleiner Perkins meeting. Borrowing a car from Rafi's parents, they headed for Sand Hill Road in Menlo Park, an affluent ghetto of tasteful office buildings set back from the road, nestled in semitropical landscaping tousled into the illusion of wildness. Billions of dollars were invested from these minicampuses every year.

For ten months, Randy Komisar, a partner at Kleiner Perkins, had been their leading cheerleader. He was, in truth, a guilty pleasure. That he had dropped into their lives was so unexpected—and so discordant with the themes of their Kickstarter campaign—they had simply kept their mouths shut about him to the outside world. They just didn't know where it was going to take them, or what it would mean.

Six months later, at the beginning of March, they would find out. They expected, as Max said, that it would simply be a matter of picking among suitors. And in truth, Diaspora got past the gatekeepers at venture capital firms with the ease of cat burglars, able to schedule meetings with a single phone call, accomplishing in a moment what most start-up firms spent months hoping to wrangle.

There was particularly good reason for optimism at Kleiner Perkins. The partner dealing directly with Diaspora, Komisar, loved the team's endearing mix: Rafi's crystal-clear-headedness, Ilya's evangelical idealism, Dan's desultory eye for style, Max's ninety-mile-per-hour enthusiasm. It was Komisar who had suggested that Bill Joy go to speak with them. And it was Komisar who had offered them the use of the incubator space at Kleiner Perkins, an opportunity that had floored them. Practically every week, either Komisar or his associate Ellen Pao spoke with the Diaspora team to get updates. There seemed to be little doubt that he was favorably inclined. During the summer, he had told them that if they ran out of money, he would pitch in to keep them going. And later, he also told Yosem that he thought Kleiner Perkins would put money into the project. He told me that he expected them to. The only question was how much.

At the formal pitch meeting, Max opened with bullet points on the desire of consumers to control their identity, and the vulnerability of Facebook in this area, as measured in dissatisfaction expressed in surveys.

Diaspora was providing an answer. There was great public enthusiasm. The genesis story showed that. It was a friendly audience; while a few questions had a skeptical tone, most of them seemed to have the purpose of prodding them to think harder and deeper. How long were people staying on the Diaspora website? How much dissatisfaction could there be with Facebook if thousands of people were joining even as they sat in the conference room? Why hadn't they taken more of the people on their waiting list?

Dan and Ilya ran through a demonstration of the site, showing a home page of a user, and a page built around President Obama. All of this was familiar to Randy and Ellen, who had been tracking the project for months. Moving to the finale, Max turned to money, a topic that had never been discussed in any depth during the chats with Randy and Ellen. His mouth felt dry as he spoke.

What were their business opportunities? They could become a trusted holder of identities, so that people surfing the web were not picked apart in the free-for-all of online tracking bugs. Diaspora did not intend to block its users from sharing with commercial operators if they chose to do so. Diaspora was set up for individuals to make better connections and have more control. They would be the "PayPal of personal data," Max said. How that would work was not defined with any precision.

He also then described a role for Diaspora as a platform for people who had created apps to sell their wares, much like the stores on Apple and Android products.

Before that happened, though, over the next year, they foresaw major growth in the number of users.

"To get there, we need more hands on deck," Max said, displaying a color-coded chart that showed about three dozen jobs that would be added.

Finally, he got to the slide.

**DIASPORA\***
**Putting users in control of their data**
**Seeking:**
**Two year, $10 million Series A**

"We have a rocket ship waiting to take off. There needs to be a few more operators," Max said.

There were no further questions. Handshakes, thanks for coming in, take care. Ellen Pao walked them to the door.

"We will call you," Pao said. "You're going to be seeing other firms?"

They were, Max said, reeling off a few names.

She wished them luck.

Afterward, they met Yosem for lunch at a restaurant in the Town & Country Village mall in Palo Alto. He was eager to hear about the session, though he was certain there was no chance at all that they had left with $10 million. The only question in his mind was how much damage they had done.

"So," he asked. "How did it go?"

"Their eyeballs did not pop out of their heads," Max said.

Dan did not agree. In fact, he felt like an idiot. The number was obscene. He saw the faces of the VCs change expression when the last slide, the one with the number, went on the screen. Until then, it seemed to him, they had been open, interested in the project and the concepts.

"I immediately saw their faces change from 'Okay, cool, I suppose' to 'That's cute; we're done here,'" Dan would say later. "There wasn't that much talk after our last slide, and we were escorted out only by Ellen." Rafi's assessment was the same. Ilya, who'd thought they were golden, was surprised by the others' reactions.

All was not lost, Yosem said.

"You should call Ellen and ask her for another chance," he advised. "Tell her, 'We are young and inexperienced, and we didn't know what we were doing.' Tell her you were just throwing a number out there because you had to, but you just want to work with Kleiner and will be happy to do so for whatever amount Kleiner thinks is reasonable."

Max nodded. He would go back to Ellen. Exhausted, they all gave themselves the afternoon to recuperate at the Sofaer house.

For pastrami, and chaos, no place in New York can match Katz's Delicatessen on the Lower East Side of Manhattan, so it took a few minutes for me to spot Ilya, even though he was the only person in the place with bright orange plastic sunglasses. He treated them as if he had spent four hundred dollars on them. Katz's was a kosher-style deli, heavy on the cured meats, knishes, and obscure soda drinks like Dr. Brown's Cel-Ray. The line went out the door, and once inside, it was self-serve at each of the big glass-covered stations, where countermen cut paper-thin slices of tongue or corned beef as samples. The tables were elbow to elbow, and when it came to finding a seat, it was every diner for him- or herself. It added a Darwinian touch to the bedlam of clattering plates and shouted announcements—"Who's brisket? You brisket or tongue? Rye or wheat?"— that a sandwich was ready.

"I love this place," Ilya said. "I only have a couple of days in New York, so I figured, bring it on."

The day before, he had been in Philadelphia to take the oath of citizenship; over the eleven years his family had been in the United States, his parents, Alexei and Inna, had already gotten theirs, and a baby sister, Maria, had it by birth. He had always intended to get it; as antiauthoritarian as he could be in his political views, he was quite sure that he viewed himself as a citizen of the United States. They might have chatted in Russian at home, but, like many children of immigrants, he wanted to be accepted in his new society. One of the final hurdles to citizenship was a civics test, he said, with questions like, What are the first ten amendments to the Constitution called, and name one civil right granted under the First Amendment.

I proposed another question.

"Whose picture is on the ten-million-dollar bill?"

"There's a ten-million-dollar bill?" he said. "No way."

I assured him there was.

"Okay," he said, half laughing, but a little uncertain. Maybe there was. "What's the picture on the ten-million-dollar bill?"

"Randy Komisar's face," I said.

He laughed. The session at Kleiner Perkins had been on Tuesday, and it was now Saturday.

"The night before was wild," Ilya said. "Yosem came in and basically tore up the proposal. He made it much better. We were so tired, after the presentation we came back to Rafi's house. Max and Rafi took naps. Dan and I stayed up, working on some problem in the code."

He did not know what to expect from the session, he said, but he wasn't worried. For all his instinctive candor, he had said nothing about his sudden departure from the project before Thanksgiving, or the circumstances of his return. Nevertheless, when asked how things were going for himself, he knew that the question was more than polite chitchat.

"I got a bike, and it's totally changed things," he said. "Things are much better now. I'm getting into things out there. Right after Thanksgiving, I got totally stressed out, and had to go home."

He was connecting with people every day. Life in San Francisco was turning out to be much better now that he was living in a real house. He promised that he would be sure to get help before things got to be too much.

A few tables away, there was a sign marking the spot where an auspicious moment in movie history was filmed.

### "WHERE HARRY MET SALLY . . . HOPE YOU HAD WHAT SHE HAD!"

Ilya wasn't familiar with the reference. The movie had come out in 1989, a year before he was born. So I annotated the joke. Sally, played by Meg Ryan, meets her platonic friend, Harry, played by Billy Crystal, for lunch at Katz's. He insists that he can always distinguish authentic sexual satisfaction in a woman from a faked performance. To prove that he absolutely cannot, Sally proceeds to provide the sounds and sights of a hearty orgasm right at the table, a performance so audible and unmistakable that it manages to override the clamor of Katz's, stunning the place into near silence.

At that moment, a waiter comes by to take the order of an older woman, who is sitting nearby and watching the writhing, moaning Sally. Looking at the waiter, then glancing back to Sally's table, the older woman eschews the menu and declares: "I'll have what she's having!"

Thus the sign at the landmark table.

Ilya howled.

"Maybe we should find out and order it, too," he said.

Yosem phoned Max a few days after the session at Kleiner Perkins to learn what he had been told by Ellen Pao in the follow-up contact. He was certain that KP was pulling back, but was not convinced they were bailing out altogether, particularly since they had spent so much time with them. It was not that the company did not want to put money into Diaspora, or was requiring a specific revenue stream—a good idea may be profitable in unexpected ways—but the more speculative the bet, the less likely they were to put down.

"I haven't called her yet," Max said. "We've been so busy with the product. Anyway, I don't think that they want to do a deal. We went for too much."

"They do want to do a deal, but not for ten million dollars," Yosem said. "It's not your fault. Don't beat yourself up because Mitch Kapor advised you."

"Actually," Max said, "he didn't advise me to ask for that."

"What do you mean he didn't advise you that?" Yosem said. Had he misunderstood how they had arrived at the figure?

"I came up with that number and told Mitch Kapor, and he didn't say anything," Max said. "So I assumed he agreed with it."

Whatever the origins of the folly, Yosem believed it essential that Max not let the connection with Kleiner Perkins get stale. "You have to call her right away," Yosem said.

Despite the goodwill they enjoyed with Komisar, which seemed genuine, their request for funding bore no relationship to any notion of where revenue might come from, or when it could be expected. Max's reading of the Netscape moment was not wrong—that company's browser, funded in part by huge investments, did, in the end, liberate the Internet—but that was not the only lesson people could draw from that episode and era. Netscape, as the avatar of the first Internet bubble, also stood for the proposition that investors could be skinned alive if they flung money at projects without seeing any plausible way to make money. That part of the historical lesson had not been absorbed into the pitch.

Their best move, Yosem believed, was the truth. They knew nothing about business, but they and others believed in the power of the idea. He strongly suspected that Kleiner Perkins would hear that sympathetically and forgive them their youth.

A few minutes later, Max reported back on his phone conversation with Ellen.

"I basically told her that we had looked at our numbers, and after reanalyzing we realized we could do it for less. She said, well, we were stunned by the amount that you asked for her, because we were willing to do a deal for $750,000."

Yosem would later say that he was shocked by what Max said next. "Max told me that he said to Ellen, 'That's too low. We'll do it for $3 million but nothing lower than that.'"

How much did the others know about this second offer? That would be a matter of dispute. Max said he had kept them informed and was following their collective will; the other three founders said they knew nothing about the $750,000 offer that Max had turned down.

What was clear to everyone was that their engagement with Kleiner Perkins had come to an end. If it was a shocking turnaround, the episode did not change their strategy. It would be several more painful months before it became clear to all of them that their charm, the raw web application they had created, and the bounty of public love were not sufficient to unlock the vaults of Silicon Valley. Everything that had happened to them so far was brand-new. Adulation or rejection, it was all in a day's work.

# CHAPTER SIXTEEN

I t was another day for blazer coats and jeans. By early April, dressing up was an established part of their routine for pitches.

Seeing them in these getups was the first clue for Mitch Kapor when the four arrived at his office. After a few preliminaries, Max pulled out his laptop, and began to run through the slides.

"Wait a minute," Kapor said. "Are you pitching me?"

Indeed they were.

Then they weren't. Kapor raised a stop sign.

"Why are you showing me your pitch?" Kapor asked.

"We were hoping you'd raise our round for us," Max said.

"I think I've overstepped my bounds here," Kapor said.

They had come to him because they were floundering, and believed that Kapor's encouragement over the months meant he was also prepared to put money into the project. After the fiasco at Kleiner Perkins at the beginning of March, the world had not come to an end. The guys continued to make presentations to the leading venture capital firms in Silicon Valley. Kapor helped them set up meetings at Greylock, Benchmark, and Google Ventures. They got an e-mail from Andreessen Horowitz inviting them in. Having watched his effort at matchmaking with Kleiner Perkins come to nothing, Yosem connected the guys with Draper Fisher Jurvetson. They also wound up pitching to Floodgate. There were others.

For the most part, the technical people at the firms seemed enthusiastic about Diaspora; it was a pretty clear demonstration that Facebook, for all its polish and smoothness of operations, did not have any digital secret sauce that it alone was able to make. What they had, of course, was scale. But even Facebook had elements of uncertainty in its business model; its ability to bring in operating revenue, as opposed to equity investors, was at that point uncertain. Regardless, the scale of its expenses was without precedent. It held the world's largest cache of photographs, and had internal standards that pictures and other elements of an individual's page—the likes, the updates, the links—be loaded at what was lightning speed. These qualities were what made the network attractive, sleek, and fleet, and thus would help bring ever more people into the Facebook corral. The basic asset of the personal data of hundreds of millions of people throbbed with intriguing commercial possibilities. Still, online advertising had a short history, making reliance on it to unlock future streams of revenue less a matter of experience and more based on faith, hope, and awe in an entity that could consistently gather so many of the planet's humans in one virtual space.

The Diaspora project also had difficulties articulating precisely how it was going to turn its relatively infinitesimal number of subscribers into money. Federation inherently meant that every user, in theory, was sovereign. The traditional stream of income from ads was unlikely if tens of thousands, or hundreds of thousands, of individuals were masters of their Diaspora domain. As the spring advanced, they began to skin back the funding they were seeking. The $10 million ask that had choked off the Kleiner Perkins presentation was removed from the slide show. Sometimes the amount was left blank; one version of it listed their request between $750,000 and $1.5 million. They would use the money to hire engineers to build out the project.

Yosem spoke privately with Randy Komisar, pleading for the crew to be allowed to return and ask for the $750,000 that Max had turned down.

"They don't have the business-side experience," Komisar told Yosem. "The very fact that you're the one asking me, and not Max or one of them, proves that."

A few days later, Max told Yosem that he was getting strange vibes from Mitch Kapor. As Yosem recalled the conversation, he suggested to Max that they ask Mitch to be the chair of the fund-raising for the first round. "That way you will get some clarity on the relationship," Yosem said he told Max. And in his view, being the chair of the round was essentially an honorary position.

For his part, Max remembered it slightly differently. As he described it, Yosem suggested that they ask Mitch to be the lead investor.

Afterward, Dan looked back at the e-mail correspondence. "Max had written to Mitch saying, 'We want to talk to you about funding.' It was ambiguous."

Whatever uncertainty there might have been on how the meeting originated, there was no doubt about what happened at it. A few minutes into it, when Kapor realized that they were asking him for money, he interrupted Max.

"I had no idea you were going to pitch me," he said. Kapor was not looking for new investments, and was in fact trying to cut back on his existing ones so that he could give more attention to the family foundation he had set up to support educational opportunities for disadvantaged children.

The meeting came to a quick, awkward close. "It's like when you know a girl as a really good friend, and you ask her on a date, and she gets way creeped out," Dan said later. "It totally creeped Mitch out."

Later, Max wrote to him, apologizing for having taken him by surprise. To dispel any sense that they had been speaking to him for months simply as a prelude to hitting on him for money, Max wrote: "We value your advice and experience first and foremost, and since we are a young team doing this for the first time, we hope we can continue to use your knowledge and experience to help us move through this crazy funding process."

In closing, Max also said, "We would love to hear any feedback you have received from other sources."

Replying, Kapor said he appreciated that Max had taken responsibility for the surprise in the meeting. He did not sugarcoat the feedback he'd gotten from venture capitalists. "There's a lot of agreement that Diaspora isn't yet in a fundable state," Kapor wrote.

It wasn't clear yet what value such a system would have for users, developers, or the groups that would host Diaspora pods, Kapor said, and no way to figure out which should get the focus. "We love the vision of a federated open social network, and we want you to win," Kapor said.

He said he could continue to be available, but did not have the time for an intensive advisory role. He needed to pull back, "though not pull away entirely," he wrote. His most important feedback, he said, was that they should pick a group—users or developers who would come to believe that Diaspora was "something they cannot live without."

That essentially was the end of their conversations with Mitch Kapor.

"We lost the connection there through ambiguity," Dan said.

Pedaling his new bike down Valencia Street in the Mission District, Ilya caught up to his friends in the pickup truck when they stopped for a traffic light.

"Give me the Bloody Mary," he shouted.

Ilya had gotten to brunch late, as the group was already paying the tab at Napper Tandy and heading for Dolores Park. It was Pride weekend in San Francisco, the annual sprawling celebration of gay humanity, and no one wanted to miss the fun in the park. So Ilya ordered his Bloody Mary to go. He asked his friend Katie Johnson to hold it in the truck until they got to the park. The rest of the group was several drinks ahead, though, and the merriment from the truck became irresistible.

Katie handed him the drink, expecting him to have a swig and give it back. Instead, sipping the Bloody Mary, steering with one hand, Ilya took off, turning a corner. Everyone in the truck bed whipped out cell phones to get video of Ilya chugging and cycling. It turned out that he was better off steering with just one hand. As he lowered the drink, he lost control of the bike and it toppled into the street, followed by the remnants of the Bloody Mary, and then Ilya. Except for one person, the people in the back of the truck roared with laughter.

"Stop! Wait," Katie hollered to the driver. "Stop."

Having invited him to the brunch, Katie felt protective. While Ilya pulled himself together, Katie ran to him.

"Are you okay? I'm so sorry. Are you okay?"

He was okay, he said.

He held up his left palm, showing a puncture.

"Awful," she said.

"No, totally, I fell in the right way," he insisted.

His orange sunglasses were intact. He unbuttoned his jacket, inspecting a scrape on his elbow, and revealing his T-shirt from *Makezine* magazine, the bible for do-it-yourselfers, with a classic message: "Void your warranty, violate a user agreement, fry a circuit, blow a fuse, poke an eye out."

The elbow scrape looked minor.

"Okay," Katie said, relieved. "We'll get you a Band-Aid."

"It sucks!" Ilya said. "I need another Bloody Mary."

His injuries weren't a big deal. He gave her a big smile and promised that he was all right. Ilya's social life, so becalmed a few months earlier that he wanted to abandon California, was billowing with new friends. The surge had started, oddly enough, with a visit to Diaspora from a group of students at Stanford, at least one of whom was keen to skin back the shimmer of hype and expose all the weak points in the project.

The class was called Ideas for a Better Internet, and the teacher, Elizabeth Stark, faintly knew the four guys from New York. The Diaspora guys knew very well who she was: Stark was a leading figure in the free-culture movement, which generally pushed for open software and free sharing of content and culture, though it distinguished itself from piracy advocates who had no regard for copyright restriction on intellectual property. Stark had cofounded the Open Video Alliance to promote technical standards that would keep the web from turning into nothing more than a one-way platform for pay television. Max and Ilya had been members of the Free Culture outpost at NYU.

A graduate of Harvard Law School, Stark had met Mark Zuckerberg when he was still a student and first building Facebook; a few years and a few hundred million users later, she gave him a heads-up about Diaspora, which led to his donation of one thousand dollars through Kickstarter. She taught at Yale and then at Stanford.

The Better Internet class studied interesting projects, met the founders, and pitched in with code or other help. Most of the students had heard of Diaspora, so Stark set up a visit to see them at Pivotal in San Francisco.

Oh, good, thought David Kettler, one of the students. I get to tell them why they suck. A little.

Kettler was a senior completing his work in a major called Symbolic Systems, which took in logic, philosophy, interface design, artificial intelligence, linguistics, and psychology. He had been working with a Stanford group, the Programmable Open Mobile Internet project, that had many goals similar to Diaspora's. And okay, he was, he knew, a little envious. Even some of the instructors regarded the four hacky kids from New York as hyped-up pretenders. They had all this attention. What else was he supposed to do besides be a little snooty?

As was often the case, it was easier being critical in the abstract, during intellectual batting practice, than in the flesh. At Pivotal, Kettler found the four Diaspora guys were friendly, welcoming, modest, and amazed at the attention they'd gotten, and showed no sense of entitlement whatsoever. Nevertheless, as soon as their presentations were finished, Kettler was ready to set them straight, cordially but firmly.

"You're focused on being your own network," Kettler said. "You should be connected with other networks."

Max explained, calmly, that they wanted to focus on building up Diaspora's own network, and then worry about opening it up to others.

"You're not going to get adoption that way," Kettler told Max. "People aren't going to go to a new system that offers no benefits other than privacy, apart from people who really care about that stuff."

He spelled out a few more particulars; his purpose was to push them, to see if they pushed back; perhaps, he acknowledged, he himself needed to be straightened out.

Ilya interrupted.

"Those are great ideas," he said, smiling. "You should come work with us and see if you can make them work. If you want to, we'd love to have you help out." Offered unconditional surrender, Kettler had no chance. That night, the Stanford students went to dinner with Ilya at an Italian restaurant.

By graduation, at Ilya's invitation, he had moved into the Hive. Even though he was working on another start-up, was in a band, and had a steady girlfriend, Kettler managed to become a regular contributor to Diaspora. Soon afterward, another member of the same class joined them in the apartment: Tony Lai, a lawyer from England who was as easygoing as

Kettler was confident. Lai had gone to Stanford for a second law degree, and was developing a tech project to democratize legal services.

At the hub of Ilya's constellation of new friends was the instructor, Stark, a tall, dark-haired, and striking presence everywhere, it seemed, that she went; she had a sprawling network of friends and younger people who looked to her for advice.

A few days after meeting the Stanford group, Ilya threw a party to celebrate his new status as a U.S. citizen. In notes inviting them, he said that he always had his parties on "Caturdays." Elizabeth arrived with friends. Ilya viewed her with awe, but she quickly became another person who crashed at the Hive rather than go back to Palo Alto after an evening in San Francisco.

Also present at the party were people who had become friends of Diaspora: Rosanna Yau, an interactive designer, who was their first employee. Sort of, since she was unpaid. All four of the Diaspora guys had interviewed her one evening at the Pivotal offices. None of them had the slightest idea what they were supposed to ask.

"What is your favorite color?" Max had asked.

"I don't have one," she said.

"What kind of a kitchen appliance would you be?" Ilya wanted to know.

Rosanna gave that a moment's thought. "A cast-iron skillet," she said.

"What is your favorite designer tool?" Dan asked.

A special pen, which was in her bag. She showed it to them. To move the conversation along, she talked about her studies at the California College of the Arts, where she had taken a graduate program in design. She was continuing research on entomophagy—the study of eating bugs for food, and how it might be made more acceptable. There was a long history of insects as part of the human diet, and it was a far more sustainable foodstuff than what was produced by modern industrial farming. The title of her thesis was "MiniLivestock: Exploring Rhetorical Methods to Promote Consuming Insects as Food." Or, as she explained it in plainer English, how branding could be used to change the perception of eating bugs.

Instantly, Ilya, who had lost some interest during the favorite-color part of the interview, was riveted. "I think this is a great idea," he said.

They were keen to bring Rosanna into the project, but the Kickstarter money had so astounded them, was so unrelated to what they wanted for themselves or any sum they had ever handled before, they could not figure out if it would be prudent to pay her from the proceeds. She started pitching in anyway, and created the Diaspora logo, a milkweed with spores flying off, as in the spreading of seeds, and remained close to the group. Ilya invited her and a roommate, Eloise Leigh, to brunch a few times. Eloise was struck by his hailstorm of ideas and his determination to keep them in sight, to make sure they were carried out. He had devised an elaborate checklist, charts for perspective, and, most basically, Post-it notes. And then there was the Jejune Institute.

All around the city, ads were pasted to lampposts and streetlights about the Jejune Institute, which invited people to learn more about the work of its laboratory. It claimed to produce things like Poly Water, a "more condensed form of water that had regenerative properties upon all organic compounds"; Aquatic Thought, "a research center dedicated to the exploration of human/dolphin interaction"; and Memory Media, "where we're able to render and record moving video images from your active memory." It invited people to visit the institute's offices in downtown San Francisco for a "FREE Induction."

In fact, it was an elaborate alternative reality game, built around a scavenger hunt that moved around the city, following clues to familiar and obscure places. It had three phases, and some people spread them over several weeks, but Ilya led a group that did it in a single weekend. The final leg required a long bike ride to the North Beach section, then up Telegraph Hill. This was no trouble for Ilya, who was an accomplished rider: when he was a student at the University of Maryland, he and a few friends bicycled home to Lower Merion Township, a 130-mile trip, sleeping overnight in a construction site. But Eloise had not been on her bike in ages and thought there was no way she could manage it.

"You can do it," Ilya insisted. "It'll be fine."

Though the hill is not terribly high, the ride up is steep, and Ilya urged them on. They kept their eyes out for a flock of red-masked parakeets, the descendants of escaped pets who lived there. At the summit of the hill stood the Coit Tower, an Art Deco monument to the days when Telegraph Hill was a relay point for passing along news of ships arriving in the bay. They were looking for the final clue in the game, but Ilya declared a pause.

"Hold up," he said.

All around them was San Francisco Bay: Alcatraz Island, innocent at a distance, and beyond it, the headlands of Marin, ripe and alive; by walking around the tower, they could see the downtown skyscrapers, and the Bay Bridge crossing to the east and coupling the city to the rest of the country. All was lit with sun.

Eloise thought Ilya was the best companion for just this kind of random, crazy adventure.

One rainy day, she and Rosanna came to the Hive, which had a big wedding-style tent permanently set up in the backyard, with a floor. A competitive ballroom dancer early in his college days, Ilya showed them basic steps, and then moved on to dances. They also did swing. On another weekend, he called Rosanna about a different matter altogether. For months, he had pestered her about making him a meal of insects, but they had not gotten around to it.

"I really want to eat bugs," he said.

Rosanna, who had been looking forward to a relaxing afternoon, tried to put him off. "I don't want to go out today," she said.

"Come on," he said. "I want you to cook bugs for me."

"Fine," she said, giving in. "I'll be there in an hour."

Rosanna had nothing planned but she brought along a few servings of mealworm larvae. The only ingredients in the larder were bread and cheese.

"Do you really want grilled cheese and mealworm sandwiches?" she asked.

"Go for it," Ilya said.

Rosanna thought it was the worst idea she'd ever had—the exoskeletons of the mealworms made them too slippery to stick to the sandwich—but Ilya raved.

"This is so good," he said.

"I don't like it," Rosanna said.

He insisted. Rosanna strongly suspected that he was being polite, but he finished every bite. And not long afterward, she was at a street food festival, and Ilya turned up to have another bug meal.

Among the people landing in his growing circle was Katie Johnson, who had gone to his aid in the Bloody Mary transport accident. She and Ilya had met when Elizabeth Stark brought her to a party at the Hive, the theme of which was billed as "We Are the Internet and We Come in Peace." Ilya was three years younger than Katie, but that was not a serious gap in San Francisco. Single people in their twenties, searching for a mission, were a social strata unto themselves. She had heard about him as a true-blue idealist from Tony Lai, the lawyer from the Stanford class who had moved into the Hive.

"What are you passionate about?" Ilya asked her not long after they met. With some people, that sort of inquiry might have come across as a line; to Katie, Ilya was the picture of earnest interest. He truly wanted to know. So she told him: she loved the outdoors, and hiking; she was interested in "collaborative consumption," a broad term covering a movement to reduce the waste of resources.

One afternoon, when a few of the crowd were having drinks, Katie mentioned the Wi-Fi password used by a friend: "FuckTomCruise."

"I told him the next time you change it, you should insert 'Yeah.' So it would be 'FuckYeahTomCruise,'" Katie said.

Ilya howled.

"Yes!" he said. "We should make a party around that. It'll be the meme: 'FuckYeahTomCruise.'"

For the party, he flipped the proper name and the epithet, and it would become "TomCruiseFuckYeah."

At night, he delivered more profound thoughts.

One evening, I had e-mailed the Diaspora Four, inviting their ideas for questions to ask Tim Berners-Lee, the creator of the World Wide Web. An interview with him was on my calendar. What to ask him? Ilya replied quickly, though he noted that he was not quite sure of Berners-Lee's role in shaping the modern world. Still, he came up with a few questions about the inventor's views on the current state of the web. His final one was not about technology, but on the pendulum of emotions that Ilya himself seemed to ride.

"Were there any truly dark times when it seemed that everything was totally doomed?" Ilya wrote. "On the other hand, were there times when everything was happening and large chunks of time were just an adrenaline rush?"

The stage was so big, so bathed in black, that the pinpoint of spotlight on the lectern where Max stood had the effect of miniaturizing him. That scale was magnified by the enormous screen behind him where his image was amplified. It was mid-April at a conference in Berlin, a long way from Silicon Valley, yet another occasion when the group—this time with Max as the representative—served as the ambassador from the world of possible, popular, and noble ideals. His subject was crowdfunding.

As it happened, crowdfunding was not on the new map of exploration being used by Diaspora in its hunt for sustenance; their trail now went through the venture capital offices of Sand Hill Road in Menlo Park. That would not have distinguished them from any of the other start-ups hoping to win the lottery. What mattered about Diaspora to the crowd at re:publica, an annual conference in Germany of two thousand bloggers and social media people, was how they had been launched by the Internet herd. They had been determined not to shift their vision or plans because of the surprising response.

"To change from that core because we had a lot of money did not make that much sense to us," Max said. "We've been very cognizant to spend the money where it makes sense—to decrease pain and increase productivity, not completely, fundamentally change why we did it."

The four, he noted, simply had wanted to create an alternative platform that developers could hack on, tinker with, modify, and build into something sturdy.

"We were doing it because we were having fun," Max said. "If it becomes a start-up or whatnot, it's still going to be why we started it. Crowdfunding in general is plugging into people's passions, and to me, successful projects are people who are honestly trying to do whatever they are trying to do. They want to do it, and they will do it whether they get the funding or not."

Other people and groups were trying to create alternative social networks, he pointed out, showing a slide naming nine such projects. The day before, at the same conference, he had heard Mitchell Baker, the chairwoman of Mozilla, the not-for-profit foundation that brought Firefox to life, talk about ways to support the open web.

Above all, she had said, Mozilla was trying to make sure that the Internet was not strictly operated by a few institutions. Centralized control might be an attractive way out of disorder, she had said, but in the long haul, it was "a recipe for potential abuse."

Mozilla and its followers were free to hold that view, Baker said, because of their status as a not-for-profit. "That we don't have the financial return as a motive, by being a nonprofit, gives us an extra ability to build in what's good for the individual and what's good for society. Firefox is a good mechanism to demonstrate that what we're talking about may be abstract ideals, but that can be made real and concrete."

When he got back to California, Max sent notes to people inside Mozilla. Baker's talk had reminded him that, he said, "More than ever, Mozilla and Diaspora are trying to solve some of the same problems."

He asked if there were opportunities at Mozilla to sponsor a project like Diaspora.

One area seemed promising: Mozilla was about to start WebFWD, a three-month "accelerator" for people using open-source tools to make the web more open. Its goal was to merge the principles of the open-source movement with the best practices of the start-up world and its emphasis on useful, competitive products. The program consisted of nine weeks of classes, plus coaching from Mozilla's technologists and the possibility for seed funding at the end of it.

Though there was enthusiasm among some people at Mozilla for bringing the Diaspora group into the inaugural class, Pascal Finette, who ran the WebFWD program for the foundation, said he did not think Diaspora's potential privacy offers alone would excite many users. His persistent question, Finette said, was, "Do you create enough user value that people will care about you?" Privacy concerns that people might entertain in their heads were not usually strong enough to take action, he believed.

Moreover, "they were looking for office space and primarily financial support," Finette recalled.

While there was a WebFWD seed fund, it was available only after a group had completed the twelve-week program. The discussions petered out, without Diaspora ever formally applying.

For his part, Max felt that people at Mozilla were not eager to take on a project with an existing high profile. He was frustrated that he had not been able to put his pitch directly to Mitchell Baker. "We suck at phone calls," he said.

# CHAPTER SEVENTEEN

**M**oney? Sitting in a café in the Mission District on a Saturday morning in May, Ilya was a shrugging study in nonchalance. "It's okay," he said. "Worst case, we ask our parents for housing money for a month."

Dressed in a surplus army fatigue jacket, a Saturday morning bed head keeping his unwashed hair pointed at spiky angles, Ilya brushed off inquiries about his health—his face seemed a touch pallid—and tucked into bacon and orange juice. Besides connecting with the people at Stanford, he had found a hacker collective in the Mission and met with an engineer at Berkeley who was cooking up a scheme to see if ratings on Twitter could be rigged with a software robot. Through Yosem, he had become involved in practical ways with Liberation Technology. Diaspora was one item on a plate full of the delicacies he relished; he had bought a bicycle, and it had changed his feelings about San Francisco. He was meeting girls. His pining for New York was easing. There was more to life than Diaspora.

Nevertheless, by May 2011, the Kickstarter money, once a pile unimaginably high to four college students, was dwindling. Not due to extravagance in their lifestyles: it was the natural erosion of a year of diligent, nonremunerative work. To begin with, the $200,000 in pledges was illusory, in that they paid $22,000 in fees to Kickstarter and $28,000 to send out the promised T-shirts and doodads to their supporters. "We should drop Diaspora and go into the T-shirt business," Rafi had said. Payroll, plus taxes, came to $114,000. Server rental costs, $8,600. A few other small expenses, and they had a little more than $25,000 left, enough for a few more months.

The experience at Kleiner Perkins had startled Ilya, who had not initially realized that the $10 million ask had been fast-acting poison, but he had enough ambivalence about the prospect of venture capital money keeping the project afloat to see a bright side to the rejection.

"Really," he said. "When I found out, I was, Yo, this is our Internet. Let's rebuild it. No corporate interests or VCs standing over my shoulder. I am pumped about that."

But how long could they keep going on housing allowances from their parents?

"Yosem, who is truly an awesome dude, came up with this idea: we can put part of it in a foundation, like Mozilla." How that might happen was not something he had given much study.

More than a year earlier, Dan fingered the delete button on his Facebook account, then kept clicking "Yes I'm sure" when the cascade of screens came along to cajole and coax him into staying. It was a moment of emancipation. Max had remained a constant Facebook user even as they worked on Diaspora at Pivotal, which the others teased him about. Dan had once been an insanely heavy user, to the point where he felt he was cutting himself off from actual people as opposed to virtual connection. Now, however, it was the other way around. After he left college and moved to San Francisco, where he had no natural social circles, the lack of a Facebook account had heightened his isolation. E-mail was not the primary connection among people his age; moreover, in a new city, the chance to follow up a social meeting with a digital shout was curtailed by his absence from Facebook, the closest thing to a cyberspace phone book. He found himself with no bridge to many people.

Worse, he was constantly hounding Ilya and Rafi to log on so he could chase down some design element, or look up someone. By April, after more than a year of absence, he caved in and signed up for another account. In a conversation with me, he seemed slightly embarrassed. "I had a month of deep thinking about my morals," Dan said. "It's a necessity."

It was only one of Dan's crises that spring. As the unsuccessful pitches piled up—it seemed to him they were up to eight by the beginning of May—he and the others struggled to think through a direction for Diaspora that made sense. At night, he would call his father, Casey, in New York to lay out the dilemmas and frustrating loop they seemed to have entered.

The momentum of their first seven months had been tremendous: from a standing start in February 2010, they had released the basic code in September and an alpha version of the site in November. But then they had stalled. The first burst of activity had been fueled by their own idealism and the global civic enthusiasm that greeted it. Now they were turning to people who used different measures to gauge its worthiness.

"It's completely free. Anybody can download it. It's great for humans on the Internet," Dan mused. "There's no opportunity for profit off the thing we are making right now. I've been trying to think of ways to make money off it but most of it doesn't jibe well with our morals." By morals, he meant their mission: they wanted to provide people with social networking tools that did not exploit their personal data, which was the basic business of Facebook, Google, and other nominally free tools.

Their pitches had been naïve. "None of us went to business school. We are right out of school. Two of us didn't graduate," Dan said. "We would start pitching our ideals and then mention a whole swath of things." The $10 million request to Kleiner Perkins "was kind of outrageous. It was kind of a joke."

Through the string of failures, Yosem Companys had remained enthusiastic. In early May, he suggested that they park the project in a foundation, a not-for-profit that would of course need funding but would have to return only a useful application to the world. There were business possibilities that could be built on such a base: jQuery, an invisible beam of code holding up many websites, was housed in a foundation called the Software Freedom Conservancy. Eben Moglen and others in the free-software movement had created the conservancy to handle some of the legal complexities associated with nonprofit companies. For that matter, Linux, the operating system that served as a prime engine of the modern Internet, was nurtured by a not-for-profit consortium. And of course, Mozilla and Firefox, the virtuous and essential spawn of Netscape, were nonprofits.

In this vision, Diaspora would be a platform owned by no one, but to which anyone could bring new applications. It was like the skateboard, the simple device capable of tricks that its early designers had never dreamed of.

"Diaspora should make it easy to embrace owning your identity online," Dan said, thinking aloud. "Completely open source to the max. We could actually get donations for that. But to get VCs to fund us now—it's basically like asking them to cut a check for a donation."

Not everyone, it turned out, was twenty-one years old and excited solely about changing the way the world communicated, regardless of the requirement that an investment have some promise of paying off. Still, if the basic Diaspora structure could not be turned into money, the applications built atop it could have logical streams. What would they be? Like the skateboard, it was hard to say. To think about something other than Diaspora when he went home at night, Dan had created an app he called "cubbies," which, in effect, made it possible to quote a picture from one site on another. It worked perfectly well with Diaspora.

The platform approach "is still doing something good for people. We can embrace our morals and still take an aggressive approach with VCs," Dan said.

His return to Facebook had convinced him that Diaspora was essential, and that nothing about the world leader was magical. It was a matter of scale. That, he knew better than most people, was an advantage that was hard to resist. He was appalled by the triviality of the Facebook news feed and the information people were handing away. Some of it was unavoidable. His social life was being held hostage.

"To get the added benefit of extended communication," he said, "I have to whore out my personal data."

# CHAPTER EIGHTEEN

As the summer of 2011 approached, interns from colleges in the East migrated to California to scratch soil in the digital vineyards. Adi Kamdar, a Yale student who had founded the university's branch of Students for Free Culture, was going to work at the Electronic Frontier Foundation in San Francisco. He was not going to be paid, and sent word to people on the West Coast who had been involved in the free-culture movement that he did not have anywhere to live. Ilya, who glancingly knew Adi because they breathed the same air at Free Culture conferences and from online chats, immediately volunteered to ask his friends. And if things didn't work out, he said, Adi could stay with him until he got settled. So Adi came from the airport to the Hive.

Almost immediately, Ilya left to go to Berlin for a week—he was on a panel at a conference—and simply told Adi that he was free to stay in his room. When Ilya got back, he set Adi up on a bean-bag couch. It wasn't luxury, and Ilya made the hospitality seem effortless. Still, Adi felt awkward about freeloading on a guy he scarcely knew.

"Can I pay you some rent?" Adi asked. "Can I at least treat you to dinner?"

"You know what," Ilya said. "You can help me throw a party."

One of Adi's closest friends from childhood and high school, a Dartmouth student named Parker Phinney, was also working in San Francisco that summer, and while he had his own digs, he had become part of the crowd at the Hive. Parker joined Adi in the party planning.

The two visitors learned the meme for the party was to be "Tom-CruiseFuckYeah."

"Where did this name come from?" Parker asked Adi in an e-mail.

"Ilya came up with, I think," Adi replied. "He was probably really high."

A network of friends and acquaintances, loosely linked by mailing lists—the "Golden Gate Fridge" and "Web Ecology" lists—received an invite from Ilya.

"On the occasion of the arrival of two very excellent human beings, Parker Phinney and Adi Kamdar if you will, we will be coming together for an evening of Tom Cruise Fuck Yeahing! A party of epic proportions, epic connections and epic libations."

It would be held "8 pm, Caturday, July 2nd, but don't show up on time cuz that ain't cool."

As Adi and Parker became more friendly with Ilya and racked up hours of bull sessions at the Hive, they developed a running joke about their new friend: the countdown until Ilya mentioned slaying the dragon. A conversation about the cute waitress at the coffee shop would meander, inevitably, to the dragon. By Ilya's lights, activists and hackers and revolutionaries, whatever their specific issues, needed to realize that they were all slaying the same powerful beast, stifler of freedom and collaboration. It had three heads, as Ilya saw it: companies that restricted the flexibility of software platforms, of which Facebook was only one example; agencies and legislators in government that made policy decisions favoring the telecom industry, the largest giver of campaign cash; and groups that kept vise-tight grips on creative work, like the Recording Industry Association of America and the Motion Picture Association of America.

"It's a triple-headed dragon," Ilya argued. Young activists and revolutionaries working on any of these fronts needed to understand that they were working in tandem, for higher purposes.

Just as he had preached to classmates hanging out in the lounges at NYU, he would hold forth after getting home from work as the group was sipping beer in the backyard, or passing around a joint, trying to get a

Hula-Hoop to begin its shimmering neon orbit around their waists as they mapped out ways to slay Ilya's dragon. Or was it dragons?

His passion about this—well, about everything—glowed from him. He was weird, at times a bit hard to follow. Also, irresistible. An instant friend to men and to women. The easy personal intimacy brought young women into his bed often enough, though he tried to avoid entanglements.

For many people, Ilya was a blur of enthusiasms, but Parker, in their bull sessions, was able to see the chain of concepts that he was linking together.

The work on Diaspora was one piece of this broader movement to create more free collaboration and communication. Parker noted that Ilya was carefully measured when he spoke in a public forum about the need for reform and the mission of Diaspora. He was conscientious about not injuring the prospects for their project with more radical propositions. (Similarly, Tony Lai, one of the roommates, noticed that although Ilya was a heavy pot smoker, he was scrupulous about doing it only in settings where the image of Diaspora would not be hurt.)

So in public, Parker saw Ilya speaking about giving people choices. At parties, in small settings, he spoke the language of revolutionary battles. He wasn't shy about using the language of an activist who hoped to fundamentally reshape structures, Parker thought. Ilya would insist that people needed to realize that movements were running in parallel: Diaspora was one of the proxies in the war for free and open software that also included the free-culture groups at universities that pushed for greater ability to draw on, not steal from, creative works, and the hacking culture that could put together a machine like the MakerBot, the 3-D printer that the NYU kids had built for under one thousand dollars, as opposed to buying one for twenty or fifty times as much.

Over the summer, Parker saw Ilya's obsession with the dragons in another context. "Epic" was one of Ilya's favorite battle cries, the term he used, for instance, not only to declare that a bit of hacking was clever and useful but also to describe the success of a crowded party that had been filled with both adventure and connections. A young man toiling away by day on lines of code might not feel that his hours were all that epic. One who conjured a three-headed dragon for an adversary would feel his work was high adventure indeed. As Parker noted, "It's much more exciting to think of yourself as a knight slaying dragons than just a kid who's trying to create competition in the social networking space."

Later in the summer, a mysterious stomach ailment made Parker quite sick. It gave him so much trouble that doctors finally decided to give him an anesthetic and explore with a scope. He needed someone to bring him home after the procedure. The first person he asked was Ilya, a kid he had really not known until a few weeks earlier.

"Absolutely," Ilya said. Ilya had made close friendship simple, almost axiomatic. He immediately put a note on his calendar. Later, Parker saw the reminder for the day of his examination. Ilya had written, "Saving Parker."

Another moment, another bellyache, elevated to epic.

In the last week of June, the Diaspora Four were looking forward to the Fourth of July weekend, with its extra day off. Ilya and Dan would be at the "TomCruiseFuckYeah" event. There were two major developments on Tuesday, June 28. That morning, Max had sealed a date to meet with another company, which was nominally just a getting-to-know you meeting. In fact, they had run through their venture capital prospects. This was an attempt at matchmaking by Randy Komisar. It had the realistic prospect of making them all decently well-off young men for their year of work. Without it, they were officially broke.

Over the long months of his engagement with Diaspora, Komisar had encouraged them to consider teaming up with Reputation.com, a company that Kleiner Perkins had backed with investment. Despite the debacle of the pitch in March, Komisar retained a soft spot for them, and continued to believe that they ought to think about joining forces with a well-funded business.

Reputation was led by Michael Fertik, a long-haired Harvard Law School graduate who had started it at age twenty-eight in 2006.

At the beginning, he had fashioned Reputation as a way to help businesses manage their images online, navigating privacy controls on social media like Facebook, and purging hurtful, hateful, calumnious statements about its subscribers from online postings.

In many cases, shoveling the Aegean stables would be quicker than cleaning up online messes. Even so, Reputation had found its way into a commercial niche that no one was yet occupying, and it had plenty of money to spend.

Fertik was about to close a deal for another $42 million in funding, and one of the items on his shopping list was a privacy-conscious social network. He was fond of saying that "data is the new oil," and that the first trillion dollars made on the web from data all went to advertisers. For the second trillion, Fertik intended to give consumers a way to either protect their data or make a profit from their share in the commodity by offering them a "data locker." As part of that plan, he had amassed commercial databases with information on millions of people.

At least in theory, a social network built on privacy was a pretty logical fit for Reputation. And for Diaspora, such a merger would give the development process a home in a business, and the guys could continue to work on it without the endless worries over raising money to keep it going and growing. Their brains and passion were evident to everyone who met them; so, too, was their lack of business experience. Fertik could teach them business.

Despite the straightforward advantages, none of the Diaspora group was enthusiastic about meeting with Reputation. They wanted to keep their own project. In this, they were no different from Mark Zuckerberg. Over and over, Zuckerberg had turned down increasingly larger offers to sell Facebook in its first few years that would have made him billions. He had no interest in surrendering control of his project, which was far more polished, far more expensive to operate, than Diaspora.

Still, it made sense to Max that they at least have a conversation with Fertik.

As it happened, by midafternoon of the twenty-eighth, such a meeting took on a new urgency. That day, Google announced that it was launching a new social network, Google+, which would give users the privacy controls that Facebook did not. Some of the corners of the technology press went beyond a simple product announcement. As the *Betabeat* blog put it: "Google Just Stole Diaspora's Thunder."

On the surface, Google offered precisely the kinds of privacy controls that the Diaspora Four had created in their alpha version. On Diaspora, the controls were called aspects—the groupings that allowed a user to segregate the tavern softball team, say, from the bosses at work. With aspects, members of a hiking club could go in one bucket, and friends from the robot project in another, and the two groups would get messages meant only for them.

In Google+, the aspects idea was called "circles," dolled up with engineering by the best software scientists money could buy. Google also offered users a way to completely download every piece of personal data they had posted on a site.

One tech blog declared that the privacy system was "taken straight from Diaspora's design documents." John Henshaw of the *Raven* blog analyzed the new network in a piece entitled "Google+ Runs Circles Around Diaspora."

"Google+ Circles works almost exactly the same way as Diaspora Aspects," Henshaw wrote. "Google took the best part of Diaspora and made it even better." He wrote that the circles feature was "blatantly ripped off," then coyly modified those words by striking through them, leaving the original words visible and replacing them with "blatantly inspired by" Diaspora.

Ilya was close to devastated, the others nearly so. Yosem initially counseled calm, saying it wasn't that serious a blow. On one level, that was true: Google's core business was auctioning off data, every second of the day, to advertisers. That was why it had $50 billion in annual revenue. Diaspora genetically could not do that; that was why it had no venture capital funding, but thousands of people donating to it. Still, the distinction—that "free" applications were actually swapped for unlimited use of personal data—seemed to have limited traction with the general public. Mom couldn't see your Friday night pictures but Budweiser could.

The term "privacy settings," as used by big centralized networks like Facebook and Google+, was "mere deception, a simple act of deliberate confusion," Eben Moglen had told Congress over the winter, before the Google+ network had been unveiled.

"These 'privacy settings' merely determine what one user can see of another user's private data. The grave, indeed fatal, design error in social networking services like Facebook isn't that Johnny can see Billy's data.

"It's that the service operator has uncontrolled access to everybody's data, regardless of the so-called 'privacy settings.'"

The Google circles scheme gave users the power to limit what everyone saw, except, of course, for Google, a business built on vacuuming up data. It was a conjurer's trick, in a way: getting the audience to watch the hand with the hat and handkerchief, while the other hand is slipping a white dove out of a hiding place. But it was a good one. And after thinking about it overnight, Yosem changed his mind.

It was a dire situation, he acknowledged.

"I've spent the past 24 hours thinking about the events that have transpired over the past couple months, and I've realized Diaspora has just entered a tricky situation that can best be described as a 'perfect storm,'" he wrote to Max. "I know this is the opposite of what I said to you over the phone when I first heard the news about Google+ yesterday."

The problem, he said, was the difference between Diaspora's public posture and its plans. At the moment, it was a social network that gave the users control over their data: "Facebook 2.0 with control as the key defining feature." That was how users, the media, and investors saw it. One day, Diaspora would be "an ecosystem of apps that builds upon user control to enable new, cool social interactions online. But this doesn't exist yet, and furthermore, no one really knows about this except us."

The arrival of Google+ had thrown a shroud over the financing prospects for Diaspora. "Some prospective investors like Tim Draper are saying Diaspora's time has passed. I've spoken to a couple of other prospective investors since yesterday, who specifically cited Google+ to reach the same conclusion."

A year earlier, Draper had been keen to meet with the Diaspora crew, but they had sensibly deferred. Now he had just looked at their slide deck, summarizing their pitch. He wasn't interested, he told them in a short, blunt e-mail earlier that week.

Facebook "is a lot stronger than they were when you guys were starting this company," Draper wrote.

Many tech writers and ordinary people who'd been waiting a year for an invitation to join Diaspora had come to something like the same conclusion, he said.

Yosem had a list of proposed responses, but the most fundamental was that they had to stop dillydallying. The waiting list of people who wanted invitations had grown to 500,000.

"One basic thing we can do is launch the product that users signed up for (i.e., 'Facebook 2.0') before Google+ came along and then announce that we have obtained 500,000+ users in a week since launch," Yosem wrote. "This will help boost us back into the news and at a more equal footing with Google+ and will prevent that we continue to be seen as a 'has been.'"

He urged them to hop off the venture capital treadmill for the moment because it was getting them nowhere. Instead, once they had 500,000 users, they could do market research on what kinds of apps would appeal to them, and get funding for those specific elements.

Finally, Yosem promised Max he would stick with them. "I realize that my email may come as a surprise to you, but I want to make sure that I'm being as clear, direct, and honest as possible about what I see as Diaspora's prospects at this crucial crossroads. And I prefer to write it as an email because it's easier to chew on, whereas over the phone it's always possible that I may forget an important detail or simply not communicate as well as I should. I want to reiterate that I'm fully committed to Diaspora and its vision, and I want to do everything possible within my means to help you succeed. I'll be with you until the end, wherever that takes us, i.e., the next Google, a spectacular failure, or something in between. But it's important for you guys to discuss these issues internally among yourselves and decide whether you agree with my assessment and want to proceed in the ways I outline above, or see things differently and prefer to do something else. Regardless of what you ultimately decide, I will be there to help as much as I can."

# CHAPTER NINETEEN

**A**t the weekend gatherings, it was inevitable that Bobby Fishkin, broad and tall, with a shock of black hair, would emerge as a person drawn to Ilya, and vice versa. Fishkin was twenty-eight and, with his father, was running Reframe It, a successful consultancy that guided groups, companies, even countries, through a decision process. It had won the 2011 McKinsey/*Harvard Business Review* Management 2.0 Challenge for the Best Process in the World to Reinvent Complex Decision Making.

Before that, he had worked on a project at Yale University on the marginalia of great authors, tracking down what had been scribbled in copies of *Hamlet* by Tolstoy, Charles Darwin, E. E. Cummings, and C. S. Lewis, among others. He created a digital annotation company with the same aims.

Technical chops were not, however, what made them so interesting to each other. Ilya gave Bobby a few verses from his aria on slaying dragons. It was brain waves at first sight.

"I've been wanting to change the world for the last ten or eleven years," Fishkin said. In college, he had visited, on what he hoped would be a romantic errand, a woman living in Barcelona. A year there had cost her $180. He had no idea such a feat was possible, but his friend had settled, for free, in an old cabaret that had been busy through the 1930s. It had since become a squat. Clothing herself was not a problem: in storage were thousands of costumes from its days as a theater.

His romantic prospects dashed quickly, Bobby continued to explore Barcelona, and those days had helped set his path for the years to come. He discovered a collective squat called La Makabra. "There were 190 acrobats occupying three or four warehouses led by a Soviet circus master. They'd created a squat university there and taught punk acrobatics to all the kids. They taught me acrobatics for the next two weeks. It was a utopian experience."

After leaving Barcelona, he started an informal association of Visionaries and Revolutionaries in the Bay Area, assembling its membership from "about 290 gurus, social geniuses, social theorists, system theorists, alternative history of the future people. Semantic web people. Free-culture people. We met for a while in an old convent. Then we went to the Yerba Buena Center for the Arts."

Ilya was intrigued. He did not put himself into any of those categories, but was more than ready to listen to people who had ideas for disruption and dragon slaying. Soon he and Bobby began hanging out.

Some Diaspora money remained in the bank, but it was all spoken for. They were technically broke and hitting up their families for support. All eyes were on Max. The guys were giving him flak for his constant assurances that "money is coming."

Max, in turn, looked to Yosem for advice. How could Max let the others feel that they had traction? Should he forward every e-mail? No, Yosem counseled. Burying the others under blizzards of e-mail about meetings, promises of meetings, hopes for promises, and so on wouldn't move the project forward. It was important, though, for the others to know what was going on, and to let them know regularly.

The assumption that Max was the CEO was a highly sensitive question with the others, Dan in particular, who had made it his postcollege goal not to work for someone else. As a practical matter, Max was listed on the incorporation papers as the CEO, and his name would appear with that title on slide pitches presented to financiers. Not only had they all seen the slide, it was agreed among them that this was simply for the purpose of having a named CEO on the slide. There seemed to be little doubt that any serious venture investor would also require that they bring in an experienced person to watch the operation.

Max had been the driving force in getting the Kickstarter campaign mounted; he had made sure that T-shirts were sent to people who had contributed money; he had set up meetings with the VCs. By agreement, when talks and presentations were given, each of them got a speaking part; all the interview and conference invitations were passed around equitably. Yet Max seemed to be acting as the leader in many important settings, and Yosem had no sense that his role was in question. The important part in the dynamic, he said, was to make sure the group had a routine for sharing information, especially if the group expanded to include Sarah Mei and Yosem, as was expected.

"The meeting should have an agenda, and everyone should have a set amount of time to talk about whatever issues they want to talk about," Yosem wrote. "This will extend to Sarah and I and anyone else who comes on board at an executive level role. The idea is to make sure the whole organization is on the same page. Right now, you guys are all friends and cofounders, so daily might be overkill. You may want to do weekly instead."

He also suggested that they scale back drastically on the amount of funds they were seeking in the short term. They had yet to collect a nickel of venture capital funding, and had not sought any public donations since their Kickstarter appeal. The tenor and morale of the other three would be propped up by any fresh investment.

"Go for the low-hanging fruit: ask for a $500K angel," Yosem advised. "Better to get the money sooner than later, lest the guys start to question their commitment to the enterprise. Remember that the longer you go without raising money, the more they might begin to question whether the idea is good enough. So the key is to try to present a continuous stream of positive news, even if that means that after raising the round, you'll have to start raising a new one."

Not long after the Fourth of July, the Diaspora Four and Yosem gathered at the Sofaer house in Palo Alto for a summit. Yosem did a flash survey of their ambitions: did they want Diaspora to be a business run for profit or as a nonprofit? This was a question he had posed to them a year earlier, when they first met in California. The world had not stood still. The rise of Google+, the continuing failure of their pitches, meant they had to do something different.

They could no longer tread water. They had to move to a public beta release—the second stage of software development, when a product is distributed in a rough-finish form so that it can be tested by many users. The idea was to get it to break: to fail faster and succeed sooner, as David Kelley, one of the tech world's leading design gurus, has put it. It was part of the maturing process. Google, in fact, had kept some of its most well-known products in beta for years, reflecting the view that applications from the web were dynamic creations, constantly evolving and being released in improved versions. This was done, in large part, to distinguish the Google age from the era when software was sold over the counter in boxes with disks in them; that software was never going to change until you bought a new box. But Google gave its software away, and kept improving it, making search and mail and photo storage ever better; with each application, as its products became more useful, the mesh of its data dragnets became finer, collecting more personal information.

For months, though, Max believed that Diaspora was never going to have the polish it needed to release even a beta version of its application unless experienced software engineers were brought on board to help. Sarah Mei, the Pivotal developer who had been a steady contributor of code and a mentor to the four, was going to be their first hire. To pay people like her, though, they needed money. They could not get to beta launch without venture funding.

Yosem retorted that to get premier venture capital firms to invest, they had to have a beta version.

In the end, they all agreed that they needed to get to a beta release as quickly as possible, and that the kinds of financing they needed was seed money from an angel investor, a different species than the normal VC. An angel was willing to take a chance on a risky venture and not demand control, but would receive a significant reward if the business took off. Yosem said he had some leads on people who might be willing to take such a gamble.

The national economy was still in poor shape that summer of 2011, three years on from the global economic decline of 2008, and there was little sign that it would improve anytime soon. They calculated that they would need $600,000 to keep the servers going and to hire experts like Sarah. Moreover, they had to raise that money by October, when it was believed that venture capital began to throttle back on investment for the rest of the year.

Yosem also recommended that they build their business selling and servicing Diaspora pods—basically, allowing an ordinary person to have control of a server, the data stored on it, and the software that ran it without necessarily understanding how it all worked, or even having one at home. This was an actual business that could grow organically from the project; it was not yet another iteration of data monetization.

As it happened there was a recent model for a business that was also a revolutionary tool: the blog. In reality, blogs were websites with simple-to-use templates for adding text or pictures. They had emerged in the 1990s as hackers developed software that would make writing on web pages nearly as simple as it was on word processors. They were among the earliest tools for ordinary people to publish words and pictures on the web. By 2003, Matt Mullenweg, a nineteen-year-old at the University of Houston, began to work on blogging code from existing software. His creation was made under free and open software principles, and when it was ready, a friend suggested a name: WordPress.

Mullenweg did not invent the blog form, but he and his friends had assembled free tools and made them part of popular culture. By the summer of 2011, WordPress was built into 14 percent of the world's websites. More than 50 million blogs were using WordPress; the parent organization reported that every month, 287 million people viewed 2.5 billion WordPress pages. Young man Mullenweg was the Gutenberg of the web.

WordPress embodied much of what Diaspora aimed for. It was free, empowering software. People owned their data. Yet it was a successful business, too, with no compromise of its values. Without paying anyone, you could get the software, create a page, and host it on your own server at WordPress.org. It also offered, by subscription, a hosting service at WordPress.com, which was the base of WordPress's financial success. Because WordPress was decentralized in its essence, no one was in charge of the blogs of the world and their contents.

The whole team believed that the basic Diaspora code should reside within a not-for-profit corporation, in keeping with the project's origins. They would keep using their existing for-profit corporation to develop other applications and build up a business hosting pods. It was no different from WordPress. That way, in good faith, they could turn again to their community of supporters to ask for additional money to keep the project going, especially if the angel investment took a while to materialize.

Afterward, Yosem reflected that the session had been the most productive, focused business meeting they'd had.

Tucked into the middle of the July Fourth weekend, the "TomCruise-FuckYeah" party grew to epic proportions. Although Parker and Adi were the guests of honor, they did a good bit of the legwork. They had worked in improv, and they sized up the backyard space. The fixed great tent had promise as a screen. Ilya found an old interview of Tom Cruise discussing his involvement with the Church of Scientology, projected it onto the roof of the tent, and ran it in a continuous loop. Dan Grippi served as DJ, playing the electronic music that he loved. Adi concocted a Top Gun cocktail, named for an early film in the Cruise oeuvre. It consisted of apple vodka, ginger beer, and champagne, cut with water. He made it by the pitcherful, adapting the recipe as the supply of ingredients shifted. Guests came in Tom Cruise–related costumes. The plan had been for Adi and Parker to cover the cost of the booze, but as the party grew in scope, Ilya insisted on chipping in. The night was a roaring, silly success.

That "Caturday" night gathering was not the end of the get-togethers that holiday weekend. On Monday, a small picnic was organized in Dolores Park by Ilya; his new roommate Tony Lai, the lawyer from London who was one of Elizabeth Stark's students; Stark; Katie Johnson, Bobby Fishkin; and a woman Bobby had been pursuing romantically for six months. By the end of the evening, Ilya and the woman had connected behind a closed door. Bobby was devastated. Ilya was flabbergasted; he had been unaware of Bobby's romantic designs.

"I had no idea," Ilya said. "Dude, I am so sorry."

Ilya assured him that he was not short on female companionship, and did not want to wreck his burgeoning friendship with Bobby. Ilya had already started seeing a bright and funny eighteen-year-old girl, recently out of high school and a tumultuous home life, whom he had met at Noisebridge, the Mission District hacker space. She had quickly and, in Ilya's view, somewhat unexpectedly moved in with him. He was uneasy about the situation, he told his roommates, but not certain how to disentangle himself without causing too much pain. Nevertheless, part of their ritual, if they were together, was a passionate kiss at 11:11 every night. Then Ilya would step out for the night.

For two straight weeks after the Fourth of July mishap, he would find an excuse to call Bobby, offering up some cheery bit of news, then suggesting that they get together. "What are you up to?" he would ask. "Let's hang out."

They would go to Mission bars like Elixir or Serendipity for Bloody Marys. If the place was hospitable, they would smoke weed; if it was not, they would go back to Bobby's place and get stoned.

The tour of Pivotal was completed. Michael Fertik, the CEO of Reputation .com, had agreed to visit the Diaspora group at the Pivotal offices. Yosem had urged Fertik to see for himself how Diaspora was different from other start-ups because of the disciplined culture at Pivotal. He saw the pairs of programmers, the orderly procession of work. He was impressed.

"This place is amazing!" Fertik said to Max. "Forget you guys, how much would it cost me to buy up this shop?" They both laughed. Two weeks earlier, at a meeting in Reputation's office in Silicon Valley, Fertik had broached an outright purchase of Disapora and merging it, and the four guys, into Reputation. Even close to broke, the guys were not that interested, but they agreed to keep talking.

Now they were gathered in a Pivotal conference room for the next stage of courtship. Fertik was a man of some charm and confidence in that charm. Just ten days earlier, venture funds had closed the deal with him for $42 million in investments in Reputation, atop earlier bets. That money was not going to stay in Fertik's pocket, and he had told Max as much. Saying that he did not like to sit at such meetings, he stood; at times, it seemed like he was running a seminar, asking questions and writing the answers on the whiteboard. That is, when he could pry a word out of them.

Before Fertik arrived, the guys had huddled about their strategy and goals. They thought Reputation would be staid, and that the culture at Pivotal had nourished them. Software engineers were hard to find. Max's favorite start-up guru, Paul Graham, had written essays about the process of "acqhire"—acquisitions for the sole purpose of acquiring the technical power of the team, not necessarily out of any interest in the product they were making. Companies with loads of cash might spend between $1 million and $5 million—most of it in stock, a kind of script with potential as a lottery ticket. Yosem cautioned them to play poker.

Fertik was direct.

"Level with me, guys. What will it take to get this deal done?" he asked.

There was silence.

Fertik began to go around the room, addressing each of them in turn about ambitions, based on his reading from the earlier meeting. "Which of you has aspirations of becoming a start-up CEO? I just want to know who I should place under my wing in the business development department," Fertik said.

He added that he could see Max as the CEO of a start-up.

"You infer correctly," Max said.

Fertik looked at Ilya. He thought he had the makings of a future chief technical officer.

"You deduce correctly," Ilya said.

The visitor admitted that he was having a harder time figuring out the ambitions of Rafi and Dan.

"I'm going back to school in September," Rafi said.

After some silence, Dan said he saw himself more on the technical side of things, but wanted to learn about the business as well.

To Yosem's eyes, all the energy in the room was with Fertik. The guys were not feeding any momentum to their suitor. More than once, Fertik said he did not want to waste their time, or his, and asked them to be direct.

Suddenly, Max said: "We'll only sell for eight million dollars."

Fertik gave a moment's consideration before he spoke.

"If that's the case, I think we should no longer continue talking," he said. "I don't want to waste my time here. There are other potential acquisitions that I could be looking at."

To this, the Diaspora four said nothing.

Where were they getting this number from? Did they have an offer for that amount? He was there in good faith; were they playing him against another company?

Again, there was silence.

Breaking the ice, Yosem explained that $8 million was a kind of rough valuation of Diaspora, based on the $750,000 venture offer that they had gotten for 10 percent of the company. That brought the theoretical value of the company to something like $8 million.

"But of course, the guys don't have a lot of experience, so they don't know what the acquisition price is," Yosem said. "In any case, what the guys care about more are the intangibles. Will they get to keep their own culture?" Meaning, would Diaspora continue to exist within the folds of Reputation?

Before the meeting, they had gone over an answer they could give to Fertik that would make clear how important it was for them to keep Diaspora going, even if housed within Reputation. The plan had been for Max to deliver it. Instead, Ilya spoke up.

"I mean, I can't speak for the rest of the guys, but I don't care about money. I care about Diaspora as an idea that will change the world. There are hundreds of thousands of people that believe in us and would be unhappy if we sold out. I just want to make sure their interests are protected."

Dan chimed in.

"We really don't care about money," he said.

The big bargaining chip was not financial: they wanted promises for the future of Diaspora if it ended up inside Reputation.

Yes, Fertik said, he wanted to keep it going, but the group was not persuaded.

The meeting ended on amicable notes. Max sent a note of apology for the awkwardness about the $8 million figure. It appeared to have rolled right off Fertik's shoulders. The Reputation team spent the weekend playing with the Diaspora site, then wrote to Max and invited them to return for another meeting. An offer was made: they would be getting salary and benefit packages worth about $200,000, and stock in Reputation, value uncertain, as it was still a private company, but it would amount to a handsome reward for their year of work.

They had, at that point, no money in the bank. The Silicon Valley culture sometimes is hailed as a perfect incarnation of Darwinian capitalism, but in truth, sheer brilliance was not enough to keep young people afloat without support from someone. The backing of the Grippi, Salzberg, Sofaer, and Zhitomirskiy families allowed their four sons to look past the offer from Reputation, jobs that paid handsomely plus stock in a company that a number of Silicon Valley's savviest investors were putting bets on. There was little enthusiasm among the Diaspora Four for this prospect. They liked Fertik, but they were not excited by Reputation, and were nowhere near ready to surrender Diaspora. Not at that price, or perhaps at any price. Though he had no formal voice or vote, Yosem concurred with their decision.

"It feels like selling out," Max said.

They thanked Fertik for his interest and declined.

# CHAPTER TWENTY

One evening in July, when Ilya seemed particularly amped up in the backyard of the Hive, Gardner watched him with some concern. A few months earlier, he'd had to grab something from Ilya's bedroom, and he was stopped short by what he saw atop the dresser: an array of prescription bottles. The Hive was a communal space, but the roommates respected one another's privacy. Gardner had never really spoken to Ilya about his sudden departure from San Francisco right after he'd moved in. There was depression involved, and now, months later, he could see that Ilya was on multiple prescriptions. A quick glance told him that these were medicines associated with psychiatric care.

That July evening, after his manic performance in the backyard had tailed off, Gardner waited for a quiet moment; he wasn't sure if the pills were all Ilya's or belonged to one of the young women he hosted.

"Hey," Gardner said. "I know you started taking some stuff for depression. I was wondering if that has ever been adjusted."

"Yeah, it has," Ilya assured him.

Onstage, Dan and Ilya were a tag team of hipster couture, Ilya in an unbuttoned flannel shirt and Dan wearing a knit watch cap. A lot of clothes for mid-July in Mexico City, but they had other things on their minds when they packed. Such as, what were they going to say?

"We're going to describe who we are," Dan told the audience. "We'll be going into what Diaspora means. Then we're going to get a little technical."

Ilya broke in.

"But don't be scared, because that will be at the end," he said.

They had a story to tell, and their audience of Free Culture apostles were keen to hear it. No one was going to demand that they show revenue streams. It was quite enough to hear about the magic of global collaboration that was driving Diaspora. They had one hundred individuals regularly contributing code. Nearly five thousand people, known as watchers, were following the daily evolution of the code on GitHub. Meetups were being held in India, Germany, Holland, Spain, and Italy. They had regular help from online figures they knew as Mr. Minus and Darth Vader.

"Mr. ZYX manages all of our translations, more than forty-five," Dan said. "It's a larger ecosystem, way bigger, than anything we could ever build by the four of us just being in an office."

Ilya, at his most vibrant, deployed his knack for explaining the complicated to discuss what they were building.

"Diaspora is a federated social network. Many of us use federated social networks today. For example, cell phone providers. I am able to call Dan, who might be on AT&T, and I'm on T-Mobile. Even though we use different providers, we're still able to communicate with one another.

"Another example is the postal service. I'm able to send letters to France," he said, even though his letter was originating in the postal system of another country.

"Or e-mail providers. I don't necessarily have to have a Gmail account; if a friend of mine has Yahoo! we can still communicate. This is something we don't have in the social networking space right now, but that's what we're building."

Dan had an example. One of Facebook's earliest, and most lucrative, successes was a game called FarmVille, which had been built by an independent company. Within two months of its creation, 10 million people were using it every month. In 2011, FarmVille would account for 12 percent of Facebook's income, $445 million.

"Say I love FarmVille. I don't, but say I did," Dan said. "I would have to have a Facebook account to use FarmVille and hook up with my friends.

That kinds of sucks for me as a user. I'm required to sign up with a single provider to fulfill my urge to plant crops and buy cows. That doesn't make sense."

It was as if videos or television shows could be seen only on certain brands of television.

"To use an application," Dan said, "you shouldn't be told—"

"Forced," Ilya said, breaking in.

"Forced to use any one single provider," Dan continued. "That's what federation gives you. It's an equal playing field."

Their audience was not techie geeks, but, like the people who had donated money a year earlier, many of those present wanted to help.

The first question was from a woman in that camp.

"Most of us have been looking forward to it since it was on Kickstarter," she said. "What can I do to contribute?"

Ilya smiled. "Dan actually talked about this very point in France."

"I gave an 'open source for nongeeks' talk in Paris a month before. God, I have this slide."

As he started hunting through the computer in search of the slide, Ilya grabbed the microphone.

"I can blabber on for a little bit," he volunteered.

"Thank you," the woman said.

"So," Ilya said. "Blah blah blah." He giggled. "There's actually several fronts. Coding is one front. User interface and feedback on that front is just phenomenal. And being sort of vocal about it."

Dan triumphantly found the slide, which turned out to be on the prosaic side, reading, "It's not all just <code/>."

"There's an impression that to contribute to open source you have to know how to code," Dan said.

"That's a fallacy," Ilya said.

"And it kind of just sucks. Even using a product as an end user—you're going to find a bug," Dan said. "No software is perfect."

"Submitting feedback," Ilya said.

"Yeah. You don't need to know code to say, 'Look, this right here is a total design failure and it's impossible for me to do what I want, right here.' Providing that feedback is superhelpful," Dan said.

He and the other builders of the project needed the outside eyes to tell them what wasn't working. There were formal channels for such suggestions. But some Diaspora users had figured out a way to get the message across: using tiny animated images, contained in files called GIFs, that packed multiple images in a single file.

"The ability to reshare a post was just put in yesterday," Dan said. "That's because a bunch of noncoders were just making a bunch of GIFs, 'where the hell is reshare?' That got a feature pushed out. Superhelpful."

The audience loved them.

Almost every month, it seemed, one or more of the group was being flown to a conference. Dan and Rafi had been to Paris and New York; Ilya had been to Berlin and Mexico. There had been trips to Venice, Providence, Austria. "Fuck it, I want to ride this thing and go to all these places," Dan had said. "This may never happen again."

A month before Mexico, Dan was in Paris when he received the invitation to speak with Ilya at the Free Culture conference in Mexico City. And before they'd even gotten to the podium there, an invitation came to the group asking for proposals to speak at a new lecture series that was being set up at the end of August at Burning Man. The speakers would get free tickets.

"Awesome," Ilya said. He, Dan, and Rafi had gone the year before, but they had not planned in time for 2011. It was the twenty-fifth anniversary of the utopian festival, and the organizers had capped ticket sales. They had been sold out at more than three hundred dollars. The lowest price through ticket scalpers was seven hundred dollars.

The guys were in demand; even without rousing success, the ideals of the project, their climb from promise to a small but real prototype, had annointed them with charisma. In Berlin a month earlier, Ilya had been halting and hesitant as part of a panel with the ungainly title of "From Society to Technology—Implementing an Interoperable, Privacy-Aware and Decentralized Social Network." He had spent part of his time onstage writing in his Maker's Notebook.

In Mexico, though, both were calm. After speaking, they dashed off a proposal for the Burning Man series, which Dan summarized as being about "the asynchronicities on the internet between the users and the service providers; the amount of stuff that we get for the amount of stuff that we give is just totally skewed." It was quickly accepted. Dan tweeted:

> looking forward to speaking about #diaspora at this year's burning man with Ilya Zhitomirskiy! see y'all out on the playa.

The tweet posted automatically to Facebook. Only one person took the time to endorse it there, using the thumbs-up symbol that is Facebook emoticon-ese for "like." That was Ilya. Both had succumbed to the siren song of Facebook.

They staggered and swaggered through the second month of their second summer in San Francisco. Essentially broke, unwilling to sell the project to the one company interested in buying it, hammered as irrelevant because of Google+, the Diaspora Four were, nevertheless, rolling in social capital. Ilya and Dan had just seen that in Mexico. As the days peeled away, Yosem suggested that they convert some of that spirit into cash. They needed to turn to both angel investors and crowdsourcing.

They needed to change not only their approach, he counseled, but their mind-sets.

That meant "thinking of ourselves as a social movement rather than tech entrepreneurs," he wrote to the team on July 24. "We currently see ourselves as tech entrepreneurs and pitch to venture capitalists as such. Venture capital pressure is driving this process: Since we want to raise venture capital money, we follow venture capitalists' advice and try to build Diaspora into something it is not.

"But think about where you came from: You didn't start as tech entrepreneurs trying to build a business model. You started as technology activists, you started with an idea that ignited a movement, and like it or not, you are now movement leaders."

It was vital, of course, that they continue to work on the project, to integrate the improvements offered by other coders, and to refine the look. But they could not function, and Diaspora could not survive, if they did not pay attention to the reality outside the screens of their laptops, or out of earshot of the acclaim of people who wanted them—or someone—to offer an alternative to Facebook. The notion of a foundation loomed, but they had done virtually nothing to move toward creating one. Such legal processes were grindingly slow. Months earlier, they had asked their lawyers in California to reorganize their corporate papers to anticipate shares in Diaspora Inc. being purchased by venture capitalists, and to make provisions for shares to be granted to advisers and people like Yosem. It had taken until July for that paperwork to be completed, and by then it was glaringly obvious that they did not have to worry too much about outside investments.

They had to stay in touch with their community of users and supporters. Their movement could not survive if they could not raise the money to keep it going until they could mount a functioning business. Social entrepreneurs "raise money from their community of zealot supporters and from wealthy patrons who share the same vision," Yosem said. "And this is what we have to do now. If you're wondering, other examples exist: Craigslist, Wikipedia, Mozilla, EFF, etc."

They had 100,000 or so people registered in the JoinDiaspora pod; about 6,000 were actively using it. There were than 500,000 more on a waiting list. Scores of pods had been set up around the world. No one had made them sign up. They believed in what Diaspora was doing, supporters in spirit and, possibly, financially. A professional fund-raiser, experienced in tapping into goodwill available for social movements, could provide a bridge of support until they were able to secure angel investments. Hiring a fund-raiser would cost money, but the arithmetic was favorable. They could pay a flat fee, betting that the proceeds of each round would easily cover the amount, or they could have a contingency arrangement, with the fund-raiser keeping a portion of the proceeds.

There was no dishonor in turning again to crowds for support. From Yosem's perspective, one of the group's bigger mistakes had been to stop fund-raising, and to go dark for weeks or months at a time with no communications with their supporters. In fairness, they were wandering in an unmapped land they had been transported to by public attention that they had not sought and by public support that they had not expected. They were kids. Everything that happened was a first in their lives. The torrent of Kickstarter money had so shocked them that Rafi had begged Max to turn it off. Kleiner Perkins had appeared to be waiting to write a check. TV trucks had parked outside the Grippi house on Long Island. How were they supposed to act? Who was going to tell them?

Having worked on the presidential campaign of General Wesley Clark, Yosem had contacts with a few people experienced in online fundraising. He advised that they speak with Peter Schurman, the first executive director of MoveOn.org, who himself had run in the Democratic primary for California governor the year before. They all agreed.

Further, Yosem said he would develop the kind of business plan that investors liked to see. He would begin to hunt around for angel investors. His official title in their paperwork was vice president for business development, but he described himself as the consigliere. Now the responsibility for raising funds was shifting from Max to him.

They had worked alongside one another hour after hour, months on end, looking like a school of fish, darting and swerving in unison. An illusion, of course, and one that could not last much longer. Any group of people in such tight intellectual and emotional space would start to get on one another's nerves. Moreover, even with their proximity, there was a great deal that they did not show one another. The decision to go back and fund-raise seemed like a rewinding to their earlier, wide-eyed days, but there were shearing forces at work.

Dan was seeing a therapist regularly. His stoic public face was the picture of hipster cool; he despaired about the project, what they were doing, what he was doing with himself. Though he was on Diaspora all day, tinkering with code and design, he did not use it except as a sandbox to see how a piece of code was behaving. Hardly anyone he knew outside geek circles was active on it. It did not serve any social function for him. He thought about leaving, but felt too bound to the others—that he owed them.

Max had a clear vision of what they would do with the first money they raised, whether it came from angel investors or from crowdfunding. It was urgent that they hire an experienced developer, he believed, and the competition for talent in Silicon Valley was stiff. Though the others did not know this yet, he was prepared to have Diaspora pay Sarah Mei of Pivotal her full salary plus a significant piece of stock—the shares that Rafi was foregoing by leaving the project to return to school—to woo her.

He had not shared the imminence of this plan with the others, at least in terms that they grasped. The weight of running Diaspora in a fashion that resembled a business had fallen, he felt with some justification, entirely on him. When he rejected Kleiner Perkins's offer of $750,000, it seemed to him that he must have told the others, or that they had to have known about it. They did not.

Rafi was heading back to school; he held little back, bluntly criticizing elements of Diaspora, puncturing some of its early claims, lacerating its pretensions.

The most public of the group, Ilya, seemed the model of transparency about his psychic aches and pains. He had confided in Mike Sofaer that he felt inadequate as a programmer—his shortcomings in that area were no secret—and had walked out the previous fall. But this summer, with the parties, the expanding circles of friends, the Hula-Hoop in the backyard of the Hive, the girls charmed by him, he had seemed lighter than air.

Yet he was very careful about the face he showed to the world. He would meet Bobby Fishkin almost every evening to smoke pot, but his Hive roommate Tony Lai had noted how very discrete Ilya was at crowded parties, making sure that no one took a picture of him with weed or alcohol. He did not want to hurt the image of the group. And the Diaspora guys knew little about his involvement with cyberactivists. One day Rafi referred to him as the prime minister of Diaspora. In a journal Ilya kept in his Maker's Notebook, he wrote that this offhand compliment had made him feel good. The other Diaspora guys knew nothing about the dresser top of prescription bottles. The young man who had learned as a high school boy how to light a stage was skilled at keeping parts of it dark.

# PART THREE

## Too Many Revolutions

# CHAPTER TWENTY-ONE

By late August, they agreed to hire Peter Schurman as a fund-raiser at a monthly fee of $5,000 from the shrinking pot of Kickstarter money. The arithmetic seemed compelling: if once a month they could get $10 donations from just 10 percent of the 500,000 people on the waiting list, they would have $50,000 every thirty days. That would come to $600,000 annually. It would let them hire engineers, get server space to accommodate the 500,000 people on the waiting list, and move Diaspora from its embryonic status into the next phase of its existence. Also, they would not be under such pressure to find outside investors, who would dilute the value of their ownership stakes and might want to exercise control.

Near the end of August, Schurman came to Pivotal to find out how he might craft the appeal. Dan was not around, and before Schurman met with the others, he sat with Max and Yosem to go over a few mechanical details, like where a "donate" button would appear in their various online presences.

There was also a huge question about what they were doing.

Could they really ask people to donate hundreds of thousands of dollars to what had been set up a year earlier as a for-profit corporation? In the summer of 2010, it had primarily been a matter of expediency: they needed a business name so they could set up a Diaspora bank account to hold the unexpected bounty from Kickstarter.

That for-profit structure had been cemented into place just a month earlier by their lawyers, who had finished revamping the articles of incorporation to create pools of stock that would be available for investors and as a lure to new employees. Of 10 million total shares, each of the four founders was allocated 1 million; Yosem was awarded 100,000; another 2 million were put into an equity incentive plan.

They could hardly go passing the hat on behalf of a for-profit corporation that they owned. The solution was to move forward, at last, with the plan to split Diaspora into two wings. A not-for-profit side would house the code that was freely available; the for-profit corporation would be the entity that conducted money-making operations, once they figured out what they would be—whether hosting pods, serving as a platform for applications, or some other business that dovetailed with their view of ethical online transactions.

It was agreed that Yosem would get the domain name, Diaspora-Foundation.org.

Soon, Ilya and Rafi came by to join the discussion, which continued over lunch at the New Delhi restaurant, a few blocks away. Revolution was in the air; could this be part of their pitch? Diaspora did not demand that its users provide real names to set up accounts, unlike Facebook and Google+.

With Diaspora, Ilya pointed out, a user could make up a pod for cats and another for his workplace.

"Cats is a silly example," Ilya said, "but there's a plethora of cases when you don't want to use real names."

He was running a chat thread on the subject, and had heard from political dissidents in the Middle East who did not want to give their actual identities in social networks, and were losing their access because the terms of service forbade pseudonyms. "A couple of activists I met are in that camp," Ilya said. "They were having an internal discussion for people banned from Google Plus—refugees online."

One of the earliest online social network projects, Second Life, was a virtual world where the entire point was to take on a new identity, using avatars. These were not transferable to mainstream networks. "A lot of them take it to social networks, but a lot of them got kicked off from Google Plus because they are not real IDs," Max said. "Even if I hated the idea of networking with avatars, because we're open source and federated, they can all start their own pod and use it. Regardless of my values, they can do so because there are no gatekeepers."

Each pod was its own dominion.

"If you live on my couch, I can kick you out," Ilya said. "But if you're in your own house, I can't kick you out."

The subject of real identities was fraught. Facebook took a rules-is-rules approach that it followed off the cliff, at times to dreadful results—for the users, not the business. After discovering that politically charged pages in Egypt were being run by people using fake names, Facebook unilaterally shut them down. It turned out that activists felt, with ample justification, that the Mubarak regime would harshly clamp down on people who were critical of the government. Thus they preferred to use fake names.

At the same time, online bullying was thought to be abetted by anonymity. In July, Randi Zuckerberg, the director of marketing for Facebook (and the sister of the founder, Mark Zuckerberg), spoke on the subject at a panel hosted by the magazine *Marie Claire*. "I think anonymity on the Internet has to go away," Zuckerberg said. "People behave a lot better when they have their real names down. . . . I think people hide behind anonymity and they feel like they can say whatever they want behind closed doors."

Eric Schmidt, the former CEO of Google, took essentially the same view, saying that he believed governments would eventually require verifiable ID in order to use the Internet. (Such a system was already in place in Korea.)

"If you are trying to commit a terrible, evil crime, it's not obvious that you should be able to do so with complete anonymity. There are no systems in our society which allow you to do that," Schmidt had said.

The Diaspora group had given this a lot of thought. For one thing, being federated meant that they could not impose an absolute requirement that only real names could be used—or that a person could have only one account, which Facebook technically did, though the restriction was easily evaded.

As Ilya noted, the prohibition on anonymity by Google and Facebook happened to work in the interests of their businesses, which involved collecting data and selling it. "Randi Zuckerberg said people are assholes online," Ilya said. "But it was not a totally valid argument."

"Diaspora is flexible enough to let you be anonymous," Max said. "There are no restrictions on having multiple accounts."

"Even if we wanted to, we couldn't do it," Ilya said. "There's people who are stalked, abused, living in repressive regimes, that need anonymous protection. I'm not sure whether valley folk don't totally understand that, but maybe they don't care because it goes against their interests."

Rafi took on the argument that banning anonymity would make for a more civil online discussion. "Some people believe people act better when they have their real names," he said. "That's wrong. There's evidence that people are more polite when they are using their real names. But being more civil doesn't mean that you're behaving better. People are less bland when anonymous."

The righteous purposes of anonymity would attract users, Ilya argued: "I think it's easier to sell people on the idea that there's Third World countries where oppression is happening that need this. That's what I'm interested in doing on my own time.

"We want to build tools to make it harder for oppressive regimes to succeed," he said. "We're comfortable pushing the big brother button."

"Big brother comes back to another thing I want to talk about, which is spinning the positive freedom thing, not negative freedom thing," Max said.

"We should fight for, not against, things. I totally agree," Ilya said. "Do we want to get on the shit list of the NSA, etc., etc., etc.? We should be focusing on the positiveness, the richness of sharing."

"We want people to be excited about Diaspora," Max said. "But messaging should be relevant to my sister. A web that's more fun is something we've been focusing on."

Since there was no central Diaspora headquarters, no ruling hierarchy that could make rules for every pod that used the software, the guys envisioned competition among the pods to offer features that attracted users or were, at least, of importance to the people running them. Among them was privacy, for which Diaspora could stake claims that neither Facebook nor Google could make.

"If the basis of your existence is advertising," Rafi said, "then privacy is inimical to profits."

"[The pods] can compete on privacy?" Schurman asked.

"Yes," Rafi said. "Some show ads, some don't."

JoinDiaspora, the pod run by the founders, did not run ads. "The entirety of our business models are not entrenched in these oversharing, privacy-eroding principles," Ilya said.

"Got it," Schurman said. "That's really key."

"And there's no implicit big brother on Diaspora," Max said.

"There's lots of little brothers," Rafi said.

"I think that's healthy," Ilya added.

Racked with fever and a cough, Dan rang Ilya. "I'm not going," he said. They were due to head off within a day on a long ride to the Nevada desert for the Burning Man festival. Besides bronchitis, Dan was wiped out from work and did not want to spend the better part of a week in the desert without showers.

By the summer of 2011, Burning Man had grown to forty-eight thousand people living in self-selected communities for a few days at the end of August. Ilya had a place in Math Camp. It had a big central tent with a few satellite tents around it, and a collection of cooking grills constantly being used. The mood there was generally more tranquil than in some of the other encampments. Now that Ilya was due to go solo—a fifty-minute talk, half speech, half questions—he fretted endlessly about delivering the message coherently to this audience.

Rafi also would not be joining them; he was back at NYU. Max was not drawn to Burning Man.

Still, Ilya did not want for company. Many friends from San Francisco made the journey to Burning Man. Aza Raskin, the design guru and entrepreneur whom he had first met at Mozilla, was staying in a camper, along with Elizabeth Stark, the law instructor and master networker, and others. Parker Phinney and Mike Sofaer drove to Nevada with Ilya and were also in Math Camp.

His Hive roommates also made the journey, and they were struck by his anxiety over the talk. Ilya rehearsed the talk multiple times with

Elizabeth, Mike, Aza, and anyone else who would listen. Aza was charmed by his nervousness, a sign of his earnestness, how seriously he was taking both his work and the moment.

The speaker series was a new addition to the events at the festival, and there was some skepticism about it.

"People taking time out of their busy Burning Man schedules to go to a talk," Gardner said. "It's Burning Man and he's freaking working. 'I've got sand in my butt crack and now I will talk to you.'"

"About slaying dragons," Tony said.

As a social experiment, Burning Man, originally an art festival, embodied the disparate ideals of radical self-reliance and radical inclusion: a city built from scratch, but with a communitarian spirit. And Ilya, it seemed to Elizabeth Stark, had lived those values. Radically self-reliant in that, having seen the problem of a mass centralization on the web, he and his partners were walking the arduous road to solving it. Radically inclusive in that he was the most generous of hosts in his very home and in his conversations. (She was tickled that one of the items on his to-do lists in the Hive was the creation of a website reflecting her own peripatetic professional life of academic appointments on both coasts, mentoring, networking, lecturing. He planned to call it "WhatTheFuck-IsElizabethStarkUpTo.com.")

Viewed from afar, the conversation in the RV might have been seen as a parody of notions fueled by the drugs that were rampantly, promiscuously present, but Bobby Fishkin—who himself had taken hallucinogenic mushrooms, and who under normal conditions was so expansive in his conversation that, as Aza Raskin said, he spoke in semicolons—was amazed that Ilya had avoided even pot or alcohol. He had wanted to be at the top of his game for the talk. Moreover, he seemed to be running on a natural fuel that needed no chemical amplification. Ilya was delighted with his delivery of the talk. "I killed," he proudly reported to Bobby, who had heard so many rehearsals, he skipped the actual event.

One night, out on the playa, the ancient dry lake bed where the festival is held, Parker and Ilya hit a party next to an "art car." Ilya wore a flying squirrel footie-pajama outfit. He jumped around the crowd, flapping his wings in front of total strangers and smiling.

"Come on!" he shouted.

Watching him, Parker saw the delight on Ilya's face as people responded. At every party, Parker thought, there's someone dancing and rocking, and you say to yourself, I wish I was having that much fun. Ilya was the guy that night.

When Katie Johnson saw him, though, there was something in his energetics that unsettled her. She had spotted him not in the squirrel outfit but wearing a wizard hat and an American flag shirt, darting around the playa. Burning Man was his natural scene, yet he seemed flighty, moving from one place to another with no apparent purpose, not really engaging with anyone. For all his antics, Katie thought, he seemed somewhat withdrawn.

At the end of the festival, Ilya, Parker, Mike Sofaer, and Elizabeth drove back to San Francisco. More and more, people sympathetic to the Diaspora cause were afraid that their moment was passing, if it had not yet already gone. It seemed to Mike and Elizabeth that they needed help navigating Silicon Valley, figuring out a way to get some cash to fuel expansion of the project. Yosem was a loyal counselor, but they didn't have any real powerhouse adviser whom they trusted.

"If we don't get VC funding, we'll try something else," Ilya said. "Diaspora is different from the usual start-up. It's going to find its own path."

# CHAPTER TWENTY-TWO

Among galvanizing forces in human affairs, two that cannot be underestimated are (1) going broke, and (2) the appearance on the horizon of a deadline. For the Diaspora group, the convergence of the two made the month of September 2011 an all-out sprint.

Since the summer, they had been running on the fumes of the hype surrounding the project and the last dollars of the Kickstarter money. Both were dwindling.

Their battle plan was to fix up Diaspora for a new release, add 2 million users, and find $2 million. And all this had to be done by November 15, their agreed-upon date to distribute a gussied-up version of Diaspora, a beta successor to the alpha they had opened up a year earlier.

None of those aims was outlandish. While Rafi, with his Buddha-like calm, was gone, Yosem was on the case full-time, joined by the fundraiser, Peter Schurman. Their questions worked like pincers that forced the lads to face problems that had long been avoided.

To become a business, or at least a sustainable social network worth joining, Diaspora had to show that people wanted to use it. In actual fact, they did. About 100,000 had signed up for the JoinDiaspora pod; other pods around the world hosted another 100,000 or so.

There was the famous list of 500,000 people waiting for invitations to JoinDiaspora, and reputable studies had shown that each new user in a social network would bring along slightly more than 3 others. So that was 2 million people, more or less.

The bigger problem was getting them to stay once they had tried it out. Only a handful of those who had received the early surge of invites and signed up were still active on Diaspora, most of whom were in Europe.

Diaspora was a neighborhood of strangers. Newcomers walked the lanes of an online world where they knew hardly anyone. They had come for the ideals of the project, fundamental among them privacy, but this turned out to mean solitude. What was the point of a social network where there was no one you knew to socialize with?

This had never been a problem for Facebook: at its beginnings in 2004, it was rolled out one school at a time. The odds were strong that anyone who joined already knew other people on the site, as existing, self-contained communities in the real world were simply being replicated and resettled online. Each additional person who joined made Facebook more useful. This was known as the network effect, a power seen vividly a century earlier with the expansion of the telephone system; as phone lines reached more users, the entire system became more valuable.

Both the telephone system and Facebook were classic centralized services. Indeed, Theodore Vail, president of American Telephone and Telegraph in the early years of the twentieth century, had persuaded government officials that a monopolistic utility was the best way to connect people with the least amount of logistical friction. (He and AT&T won the day, and the phone monopoly, once it was consolidated, remained intact for most of the next eight decades.)

Suppose JoinDiaspora actually, suddenly, brought in 2 million people, Yosem suggested. Perhaps the lack of an organic community—a school, a business, a church, whatever—could be turned into an advantage. Diaspora could be positioned as a place to meet new people. Users could identify interests with hashtags, and a welcoming committee could greet the newbies. They'd help steer the lacrosse players and the Asian food groupies to the right places online.

"Having them connect with a random sample of existing users would be beneficial, and having the random sample know that a new user has joined so they can welcome the user," he suggested.

The idea got refined. They solicited members of the hard-core sup-porters to serve as a welcoming committee. Quickly, a team was ready.

Yosem had also taken concrete surveys of who was using or had tried Diaspora, and what they wanted. Mostly, they were guys between eigh-teen and thirty-four. Mostly, they lived in Europe or the United States.

Time and again, Yosem reported, those users and others said they wanted to be able to own their own pods. He had surveyed twenty-five thousand online social network users and found that more than half were interested in a one-click installation of a personal pod, hosting their own Diaspora service, and were willing to pay at least eight dollars a month for someone else to take care of the details. "It is a good revenue generator if we can pull it off," Yosem said.

Projecting the results of his survey and other market research, the potential market was 20 million users, $1.9 billion in annual revenue, according to the plan.

The Diaspora software would always be free. People could download it for free and set up their own pods, or they could pay Diaspora to do that. This paralleled WordPress, which provided its blogging software for free and offered hosting services at a cost. It was a pathway to doing well and doing good. Hosting was the business.

To create these privacy pods would take another year, or at least well into 2012, the group had estimated. Before then, they would build up the users on their home pod, JoinDiaspora, to 2 million by inviting the 500,000 people on the waiting list to join. Until they got going with the business, though, they had to raise money to keep themselves and their infrastructure alive. One source of funds, though an uncertain one, was people who believed in their mission and would donate. After Peter Schurman cranked up the e-mail solicitations, they'd have an idea how viable that was. Another possibility was angel investors interested in the pod business. Yosem contacted friends at Turms Advisors, a group in Sin-gapore that found opportunities for Asian investors. The Turms people were confident, Yosem reported, that they would be able to put together a $1 million angel round without much difficulty; Asian investors did not have the same investment opportunities in Silicon Valley. Before that money would be committed, though, the Turms representatives wanted to meet the team and see how the developers worked.

They had also agreed in principle to provide fifty thousand dollars for immediate expenses. Max had a clear idea of how he was going to spend that money. He, it turned out, was the only one of the group who did.

Sarah Mei was not the kind of person Max wanted to hire as soon as they got money; she was precisely the person. Like so much of the bounty visited on Disapora, she seemed to have turned up out of nowhere during their first summer in San Francisco, a name they knew initially from their computer screens, and then in their daily lives when she revealed that she was working for Pivotal on the same sprawling floor. An expert in Ruby on Rails, the language that they were using to build Diaspora, she was also the mother of two young children, and she admired the idealism of the four young men.

The entire group had come to rely on her judgment and expertise. Max believed that to meet the mid-November beta release, they needed to bring Sarah on board immediately as the chief technical officer. In the materials prepared for the angel investors, Sarah would be paid essentially the same salary she was making at Pivotal, $130,000, and Max had also promised that she would receive the founder's stock that Rafi had forfeited by leaving the project. That came to a 7.5 percent stake.

This shook Yosem. Every dollar of the first investment money they had lined up would go to one person. The proposed arrangement also disturbed the Singapore investors, who wanted Yosem to explain why she was receiving such a significant equity share and salary. It was customary for people in start-ups to receive one or the other, not both.

As it happened, Diaspora had also recently retained as an adviser Fadi Bishara, a Silicon Valley matchmaker who they hoped would connect them with investors. He, too, would be getting shares of stock for his work. Ilya had met him in Atherton, California, at the Blackbox Mansion, a kind of fancy dormitory set up by Bishara and others to host hardworking young geeks who were not too fussy about bunk beds. Bishara also had been a headhunter for Silicon Valley companies.

Yosem turned to Bishara for advice on the proposed compensation for Sarah Mei. He responded that it was "absolutely too high." Embryonic start-ups like Diaspora virtually never had the cash to pay market rate;

instead, they offered stakes in the company. "No start-up at this stage pays a market rate for anyone to join," Bishara said.

Yosem spoke with Max about the feedback and told him that not only was their local adviser opposed to the terms, the Singapore advisers had said "no angel investor in his right mind would invest" if that one salary would consume so much of it. The whole arrangement put too much leverage in the hands of a single person.

To Max, that line of thinking was obtuse. The supply of talented software engineers could not keep up with Silicon Valley's demand for their skills. A less-settled person might be able to take a gamble on the future of Diaspora, but Sarah, the mother of two children, living in one of the most expensive cities in the United States, could not walk away from a salary. She had already given them hundreds of hours of time on weekends and at night.

"Sarah will get what I said, and if the investors don't like it, they can go screw themselves," Max told Yosem. "I'm not going to let people who are just investment bankers and headhunters make product decisions for me. I want to get the best talent, and to get the best talent, you have to pay the best. So if that means I have to pay them the full salaries they could get elsewhere, plus stock, so be it. I'm not going to have some external person say what I can do and what I can't do with my firm."

Yosem urged Max to build a persuasive case for the package. In more than a year and a half of comity, it was the first time Yosem had seen or heard him lose his temper.

The first phase of Peter Schurman's fund-raising campaign began with a declaration of victory, posted on their blog and written in fund-raiser-ese: "You're already changing the world, and don't stop!"

"We can't help but be pleased with the impact our work has had on two of the biggest developments," he wrote. "We're proud that Google+ imitated one of our core features, aspects, with their circles. And now Facebook is at last moving in the right direction with user control over privacy, a move spurred not just by Google+, but more fundamentally by you and tens of thousands of community members, as well as hundreds of thousands of people who've lined up to try Diaspora—that is, by all of us who've stood up to say 'there has to be a better way.' We're making a difference already."

Schurman also began composing e-mails to bring the waiting-list community up-to-date on the project, prior to hitting them up for donations. Haggling began. Ilya weighed in first. The young man who had spent the year hovering at the distant edges of various world uprisings did not want to advertise dissent as a capability of Diaspora. He objected to a line in the draft that said: "You can be a revolutionary, and connect with other revolutionaries, and if the government where your pod is hosted doesn't like it, you can organize your revolution on another pod somewhere else."

Ilya said that while he wholeheartedly agreed with the sentiment personally, to look too much like upstarts "may not be helpful for the project."

Max weighed in with a sensible concern: the messages, while directed to people on the waiting list, should not neglect their Kickstarter backers, who had already gotten invitations.

One more thing from Ilya. "Data deletion" had been a goal of the project, and one that had been mentioned in early discussions of what Diaspora would be like. In the draft, Schurman had written: "You can unplug any time, and if you do your information stays with you, not with us."

They actually couldn't promise that, Ilya pointed out. A post on Diaspora was like e-mail: people who wrote and regretted an e-mail could delete it from their own computers and their accounts, but they had no ability to make the people on the other end—the ones who received the message—delete it. And just as in e-mail, the recipient had a copy of Diaspora posts. So that claim had to go. The subject weighed on Ilya's mind.

Yosem also wanted to show people that they had been careful with the money already donated and, as proof, proposed that they include a link to a profit-and-loss statement that would be comprehensible to anyone.

With their short-term goal of raising public donations, not investments, they had to show that a fence existed between what they hoped to make money off and what they planned to give away for free. The idea had been in the air for months, and Yosem and Max agreed it was time to move ahead with it. From his research, Yosem believed that a "foundation" could exist without an official blessing from government authorities as long as donations were not tax deductible. What precisely would belong to the foundation was uncertain, as Diaspora's lawyers did not know which pieces of intellectual property would be of interest to investors.

Nevertheless, they needed to fence off in some public way the entity that was soliciting donations. On September 16, Yosem registered the domain name, DiasporaFoundation.org.

As soon as Peter sent out the first e-mail blast, the volume of inbound e-mails doubled to five hundred. Yosem intended to recruit volunteers to answer them, but wound up replying to virtually every one himself. Their Twitter followers jumped by more than a thousand in the first week after the e-mails went out. He arranged for Diaspora to join a "Digital Due Process" coalition formed by the Electronic Frontier Foundation to lobby for reforms to laws on the seizure of personal communications over the Internet. Among the early members were Dropbox and Apple.

A few of the incoming e-mails had to be shared with everyone in the group. "Hi," one of these began. "My name is Joey Grassia and I work on the advertising team at Facebook.

"I would like to speak with you about some fabulous marketing opportunities between Diaspora and Facebook."

On September 28, they landed a major publicity coup: National Public Radio was doing a story on anonymity on the Internet, quoting Eric Schmidt, the CEO of Google, on why Facebook had the right policy in place by requiring real names.

To rebut him, NPR spoke to a blogger who said she was very happy to be on a new social network that permitted anonymity, something called Diaspora, which was "still in the process of rolling out," said Martin Kaste, the NPR reporter.

He continued: "At Diaspora headquarters in San Francisco, cofounder Max Salzberg says this network does not insist on real names."

Then Max was heard: "Certainly, with Diaspora, it's not a requirement and, in fact, it's not even something we can enforce because Diaspora is open source."

Max was gifted with being able to state both sides of an argument, and the reporter noted that he agreed with Google's Schmidt "a little bit." But he urged an appreciation for pseudonyms, saying that they could be well-constructed identities.

"And it's still, like, an authentic personality they have," he said.

Salzberg duking it out with Schmidt, and the kid got the last word.

Toward the end of the month, Ilya told Bobby he had something that he wanted to give him. It was a gadget for organizing sticky notes. Bobby had been talking about his sticky notes for months, and how they piled up with incomplete to-do lists and ideas. Fishkin was charmed by the aptness of the gift and how utterly inattentive Ilya had been.

"Ilya, the sticky notes that I have are virtual," Bobby said. "They are not on paper."

Unlike Ilya's sticky note collection, Fishkin's were generated by a software program and were visible only when the computer was on. "I have two thousand virtual sticky notes on my machine," Bobby said.

"They're not physical notes?" Ilya said, astonished.

Fishkin shook his head.

"Even so, I want you to keep your revolutions organized. This is my wish for you. I want you to keep organized over the years to come," Ilya said.

He had stocked up on the organizers, buying about twenty, and presented them to people like Elizabeth Stark and his roommates in the Hive. Ilya kept surprising them.

In late September, the two advisers from Singapore, Antonio Lopez Abello and Emilio Manso-Salinas, sent word that they planned to visit as part of a due diligence inquiry into Diaspora. Sarah Mei was curious, and asked Yosem what it would amount to.

It was a particularly relevant question for her. After many months of discussion about her taking the post of chief technical officer, she and Max expected that she would be starting in mid-October. It was almost time for her to give notice to Pivotal.

The advisers wanted to meet the team, Yosem explained, and go over the financials.

"Once we have that, we (Antonio, Emilio, and I) will be pitching to investors non-stop until we close the first $1 million round," Yosem wrote. Late on the evening of Monday, the twenty-sixth, Yosem got in touch with the men in Singapore. Sarah had given her two-week notice that day; that meant she would be able to start in mid-October with Diaspora. They would need her every minute from then until the beta launch, which was scheduled for mid-November. The logistics of getting the fifty thousand dollars in tide-over money were critical. Sarah would need a paycheck by the beginning of November.

"Wow," Lopez Abello replied. "No pressure."

A few minutes later, Yosem wrote back with an update. He had spoken with Max, and there had been a change in Sarah Mei's plans. Pivotal was sending her to a conference, so she would delay her departure for a few weeks.

# CHAPTER TWENTY-THREE

In the early evening of the first Saturday in October, Yosem's phone lit up with a call from Dan. No one kept banker's hours, and the surging frenzy of the last month had found its way into just about every waking hour, but still, it was a surprise for Dan to be ringing him on the weekend. Dan normally had his nose in the code; Yosem channeled his communications through Max. There was no need to be pinging all of them every four minutes with the latest wisp of administrative business.

Yosem, for his part, had not been so busy since falling ill two years earlier. Still on a cocktail of hormones that was constantly being adjusted, he had plowed past his own fatigue, exhilarated that, somehow, Diaspora was finally moving toward a moment of truth.

"Yosem," Dan said, "why are we saying that we're a foundation?"

The question startled Yosem. So did the edge in Dan's voice. Two days earlier, Yosem made DiasporaFoundation.org a gateway into the project. After talking about this on and off for more than a year, it was time to get it done as they were on the verge of asking for donations.

"I had been told by Max that it was okay with you," Yosem said.

"We're a for-profit entity. We're a corporation. This is not okay at all," Dan replied. "That's like lying to people. We're not a foundation."

Yosem was engaged to be married to Emily Greenberg, a medical student whose mother was a lawyer for nonprofits. He had consulted her, he told Dan, and learned that because they were not selling any products for money, they could take donations and consider themselves a private foundation. They would in time have to set up a not-for-profit corporation, and shift the free-software code there—this was the approach taken by WordPress—but they had been advised not to change anything until investors figured out what parts of the intellectual property they wanted.

It made a certain amount of slapdash sense to Dan once Yosem had explained it. "I discussed this all with Max," Yosem said. "He said you had all agreed to this."

"I never heard any of this," Dan said, "and I never agreed to any of it. And I'm sure Ilya has never heard about it."

There was something else on Dan's mind. How had it happened that Max ended up in the NPR story about anonymity?

"Why haven't I been hearing about any of those interview things?" Dan asked.

Yosem was rocked.

"I've been sending all those things to you guys. Max was supposed to be telling all of you this."

A bucket of frustration tipped over. Dan told Yosem that he had run out of desire to keep pushing Diaspora up the hill. He and Ilya were going to give it until Christmas, then leave.

"Max treats us like employees," Dan said. "Why is our business plan involving a super-senior engineer when we can have, like, cheaper-paid people, interns? People were saying, 'You need a few out-of-college-kids.' You and Max keep saying no."

"I haven't been saying no," Yosem said. "That's what I've been trying to convince Max. But he keeps saying, 'We need this.' I thought he meant the team had talked about this and thought it through."

"Absolutely not," Dan said.

How could they have lived such separate realities? Yosem asked Dan if he knew the terms that Max had offered Sarah: her full salary of $130,000, and the 7.5 percent of the company's stock, the portion that had been allocated to Rafi but had not vested before he went back to school.

"What?" Dan exploded. In the first place, they shouldn't be giving her that much of the company. And second, did they even have the money to pay her?

The investment group from Singapore, due in town the following week, was going to write them a check for fifty thousand dollars. "Basically, that money will go to pay to Sarah," Yosem said. "I thought you knew about this." Although, he cautioned, that money was not yet in hand.

"Sarah has kids," Dan said. "She needs insurance. She's not like us."

In his role as consigliere/mentor/adviser, Yosem spent hours talking to Max, assuming that he conveyed the essentials to the rest of the team.

"Max would say, 'We want to do this,' and I assumed 'we' meant the team," Yosem said.

"You thought you were talking to all of us," Dan said.

In Dan's view, Max was CEO only for the purpose of having someone with that title on the slide shows that they presented to the VCs. Yet in reality, the others seemed to have given him the authority to function as CEO, if only because he had energy for taking on thankless tasks—like making sure thousands of T-shirts got mailed to their Kickstarter contributors from which the others would avert their eyes.

Yosem and Dan then reviewed the aftermath of the Kleiner Perkins fiasco. Dan recalled that Max had taken a call from Ellen Pao while they were all in the Pivotal office. Afterward, he reported that Kleiner was not going to fund them.

"You know that they made an offer?" Yosem said.

All Dan knew, he said, was that they had been turned down. He listened to Yosem unspool the narrative of what happened after the meeting: that Yosem had urged Max to go back and withdraw the $10 million request, and to write it off to them being college kids with no business experience; that Kleiner Perkins had instead made a far more realistic offer of $750,000, and that it had been turned down by Max, who insisted that they needed $2 million to $3 million.

It was news to him, Dan said.

They spoke for more than an hour, each revelation infuriating the other in some form: for Dan, it was all the business decisions that he believed Max had made without consultation, and for Yosem, it was that so little of his conversations with Max were making it to the other members of the group.

Dan signed off the call with Yosem and immediately contacted Ilya, who soon was on the phone with Yosem to get this news himself. After a few more phone calls, they had decided that Max would be told on Monday that he was not the CEO. And that they could not hire Sarah Mei.

The conversations continued over e-mail, where Yosem went through every business item with Ilya and Dan. It was almost impossible to believe that only then were they engaged with basic decisions about their project and its business. Yet it was the same blithe naïveté that had allowed them to take on a task as formidable as reorienting the web. Some of this had already been packed into e-mails that had been circulated containing the business plan, but as far as the others were concerned, they had not talked it over.

"My understanding was that the board had decided to hire Sarah and had written up her job offer," Yosem wrote.

"Re: Sarah—none of that was discussed between the board," Dan wrote.

Ilya weighed in with an endorsement of Sarah—that she would be "an excellent person to join us"—but added, "Unfortunately, there was no codified board meeting nor a job offer that I know about."

Both Dan and Ilya added that they had no idea about the stock award for Sarah. "Let's not do that again," Dan wrote.

"100% with Dan on this one," Ilya wrote.

As far as having structured officers, Yosem described a few titles, then addressed the elephant in the room. "My understanding was that you guys had agreed that the CEO is Max. Is this symbolic or operational?" he wrote. "We don't really need a CEO for the time being, except for symbolic, external investor purposes."

Dan replied: "When we agreed that Max was CEO, it was to put a title on a slide for a deck. At the time, all I ever agreed to was the symbolic nature of the title. Obviously, this has not been the case. We need to discuss the details of this amongst the board."

"+1," Ilya wrote.

As they spoke through the evening, Max was part of another conversational thread about drafts of a fund-raising letter that was due to go out in the next few days. The others did not reveal what they were speaking about, so he knew nothing of the mutiny that was roiling.

Yosem discussed with Dan and Ilya the job of "virtual CEO," which would involve a mentor-type figure advising them and being their front person at pitches with investors. Randy Komisar had played that role for a number of companies before joining the investment bank. It was "purely symbolic," Yosem said, but it "means that I can meet with investors and coach you guys as founders into the CEO role. But Max didn't feel comfortable doing that, which is ok. One thing I could do to help in this regard is hold regular seminars for you guys on how to run a startup company, like a mini-MBA, as I do for my Stanford engineering/entrepreneurship students."

"I would definitely love regular mini-MBA seminars (business is something I know very little about, but would love to know more)," Ilya wrote.

They would hold a board meeting Monday morning to tell Max that he was not the CEO. After that, they would tell Sarah Mei that her hiring was going to be put off. Yosem sent around an e-mail at nine-thirty that night to the group; it was entirely for Max's benefit.

"Dan called me a while ago, and we spoke for a bit," Yosem wrote. "The gist of it is that it appears that we're experiencing some communication issues.

"To avoid them in the future, we should hold a weekly meeting where, during the first half, I can keep you abreast of our progress on the business front, and during the second half, say over lunch, we can have Peter and Sarah join us."

After midnight, Dan sent Yosem an e-mail. "Just wanted to say thanks again for taking the time to talk earlier tonight," he wrote. "You've got some great insight, as always."

They had scarcely ever held a board meeting, and on Monday morning, no one was quite sure if Rafi was even still a member. Had he automatically resigned when he left the project to go back to school? It was spelled out somewhere in their incorporation paperwork, but they did not have time to figure that out. To be certain that he was notified of the meeting, Dan sent him a note. Did he want to join them by phone around eleven or eleven-thirty Pacific time? Not possible: Rafi was in class then. But he wanted the minutes.

The three of them gathered in a conference room.

"You're not calling yourself the CEO anymore," Dan said. "You're not the CEO. That was a title we put on a slide. We are equals in this."

He went on to recite the developments that he and Ilya had learned about only over the weekend: the hiring of Sarah Mei, and the awarding of 7.5 percent of the company. That, Max said, had been part of the plan all along. It was in the business plan they all had been sent.

"We never made a decision to make a formal job offer," Dan said.

Ilya said that he thought Sarah Mei was great, but that he had known nothing of the plan to give her Rafi's stock.

"She's giving her notice today," Max said.

Dan swore. They were paying themselves virtually nothing, and any new money was committed to one person making $130,000, who was also getting a hunk of the company? It made no sense.

Max insisted that was the price for Silicon Valley talent, which was what they needed to move forward.

What about the Kleiner Perkins offer? How come he had not told them about it?

"I went back to you guys, and you said no," Max said. "Don't blame me. That was what you guys wanted."

Dan and Ilya said they had known nothing of it. Max insisted that he had told them. No, said Dan.

"That would have changed everything," Dan said.

Going forward, the three of them would function as co-CEOs. It was a silly notion; they had not figured out how they would divide the tasks. But in truth, it was a holding position. Yosem was waiting in the wings.

How often did start-ups fail? Depending on who was counting, nine out of ten, or six out of ten, or some other majority fraction. In the tech world, failure was not a failing; it was part of the culture of trying. As Drew Houston, who started Dropbox at age twenty-four after he got frustrated on a bus ride from New York to Boston because he had forgotten his thumb drive, noted: "Bill Gates's first company made software for traffic lights. Steve Jobs's first company made plastic whistles that let you make free phone calls. Both failed, but it's hard to imagine they were too upset about it." The Bay Area teemed with small groups, two or three people who had an idea for something. Even the most devoted band of brothers, or pals from the computer club room, would succumb to the tensions of working side by side for hours without end. Ilya thought it was fortunate that they had not all ended up living together. Being able to tolerate partners took work.

Diaspora did not fall under the standard rubric for evaluating startups; it had no investors waiting for returns or for a big exit buyout. They were not subject to the discipline of venture capital, for good or ill. It was more than a year and a half since they'd heard Eben Moglen's speech and run away to join the circus. For nearly all that time, Max, by default and by aspiration, had been the person who took care of the business side—whatever it was or might be—in between coding with the others. Or, in Ilya's case, as he learned to code. In truth, Max had been stretched in a million directions. As he saw it, now he was being blamed for them not being a functioning business, for not telling them about things that they did not want to know about. Dan was ready to kill him, Max thought.

When the meeting ended, Max stalked out of the conference, over to the table space that Pivotal had given them. He slammed down some papers, put on his headphones. He was done speaking to people for the day.

Dan turned into the aisle and headed across the room to Sarah Mei's desk. "I'm sorry," he said. "Don't give notice."

They were not ready to move ahead with her hiring, for reasons that had nothing to do with her, he told her. They thought she was awesome. Sarah took the news calmly.

During the evening, Yosem noted in an e-mail to Dan and Ilya that Max had recently offered him 5 percent of the stock. "I didn't mention this in front of Max at today's meeting, however, because I assumed that Max had not discussed this with the board either," Yosem wrote. "As I see all four of you as my friends, please decide my stock compensation based on what you think is fair given my contributions to Diaspora during the time I've been working with you. As you know, I've worked on Diaspora, not because I care about compensation, but rather because I believe in what we're doing and really enjoy working with you."

The same, in fact, could probably be said for Sarah Mei, who had found time in her life to improve their code without remuneration. Later that night, Yosem dropped a note to her. The Singapore advisers had postponed their trip—one of them was in the middle of a divorce—and the funds were not in hand. "I feel bad," he wrote, "because I know you were excited about coming on board, just as we were excited to have you join us full-time."

She replied with equanimity. "Pivotal is a pretty nice place to be, all things considered," she said.

A good thing: the next day, after checking with the lawyers who were working on contingency for Diaspora, Max and Yosem separately learned that Rafi's stock did not automatically revert to the pool because he had left as an employee. While he was still on the board, his stock would continue to vest. They could make other arrangements, but as of the first week of October, his status had not changed.

What that meant is that not only could they not have given Sarah Mei the first fifty thousand dollars from the angel investment advisers, they could not have given her the 7.5 percent stock, either. They had neither the stock nor the money in hand.

The next day, a reporter from *TheStreet,* a financial news blog, asked to do an interview for a story on how competitors to Facebook were "chipping away" at its hold. He noted that Diaspora was a potential rival that was generating a lot of buzz.

The interviews were a particularly tender point, as Max had been on NPR just the week before. They had agreed to return to a strict rotation for dealing with press inquiries. "I believe it's Max's turn," Yosem wrote. "Max, you want to take this one?"

Max tersely declined.

With everyone in charge, and no one, they decided to meet on Wednesday evening and discuss the leadership situation over dinner. They planned to go to a German restaurant, Schmidt's, but instead coded through dinner hour. After waiting three hours, Yosem headed back to Palo Alto, and the Diaspora crew went to a Meetup at Shotwell's, a bar in the Mission. Dan called Yosem from outside and said that Max was making him and Ilya uncomfortable.

"This has got to stop," Dan said to him. "You have to put your foot down."

For the rest of the night, Yosem got legal advice. Max had to write a single sentence resigning the title of CEO, since that was how he was listed on the incorporation papers. If he refused, the other board members could strip—"destitute" was the legal term—him of the title. He suggested that they engage a virtual CEO, a position he volunteered to fill, or to help them fill. "The point is the business really needs someone with business experience to serve in that role, whether it's me or someone else," he said.

"We've tried on the job training, along with my pretty much coaching Max for the CEO role over the course of a year, and there have been multiple communication issues, few results on the financing front, and dissatisfaction all around," Yosem wrote in an e-mail that he sent to Dan at four in the morning. "This is really not Max's fault. Imagine if I had asked and tried to run product development."

As virtual CEO, Yosem said, he would serve at the pleasure of the board, which would have the final say on all decisions. The board meetings would be, in essence, training sessions. They would become "Q&A sessions, where I explain the logic of business decision-making, make recommendations, and the board votes and makes decisions."

He ended on an emphatic note. "You told me last night to put my foot down, and this is where I'm going to put my foot down. Enough is enough, and you're right, the time has come."

Except, Dan decided in the light of Thursday, the time had not come. He had not talked it over enough with Ilya and Rafi. They'd have a board meeting on the following Monday, and everyone would be prepared.

At the end of the first full week of turmoil, Ilya needed a clear break.

"There's a secret rave being held on a cruise ship," Bobby Fishkin announced.

It sounded perfect to Ilya.

A friend of a friend of Fishkin's had purchased the thirteenth largest cruise ship in the world, the first oceangoing passenger liner built by Germany after World War II, with berths for three thousand people. The ship had long since run out its commercial life and, after a period when it was a kind of dry-docked hotel for a cult, was bought by a ship aficionado named Chris Wilson for one dollar. In theory, he could have sold it as scrap metal for a handsome price. Instead, Wilson, a man of quite ordinary means, set about restoring the ship, now known as the *Aurora*, and had it towed to Pier 38 in San Francisco Bay. He discovered that it was not welcome, a subject of energetic enforcement of maritime and port codes by harbor authorities who made frequent inspections. Its main sin appeared to be that it was an unsightly presence in the venue chosen for the 2013 America's Cup, the international yachting competition that had been brought to San Francisco by Larry Ellison, the founder of the software company Oracle, and one of the world's wealthiest people.

From time to time, Fishkin explained to Ilya, Wilson quietly rented out the space to help pay his storage bills.

They made their way down to Pier 38, along the darkened waterfront, and boarded the *Aurora*. About three or four hundred people had turned up, and they were all collected in a few spaces on the ship, waves of electronic music pulsing across the decks. Officially, no drugs or alcohol were permitted. Ecstasy would not have been unusual, though.

Whatever was going in the actual rave was not on the agenda of Bobby and Ilya. In fact, Ilya had once told Max that his ideal girl was a raver who did not do drugs, an esoteric combination.

Through Bobby's friend, they were given a tour of the boat, which seemed like a skyscraper lying on its side. It had ten stories. Some three thousand place settings were stashed away. Down they wandered to the lower decks, to engine rooms, to the galleys and any other cranny. So big was the ship that the swaying ravers were unseen and unheard.

"We could bring two thousand people here, and it would still feel empty," Bobby said.

Ilya's eyes lit.

"What if we got one thousand hackers to hack the boat?" he asked.

# CHAPTER TWENTY-FOUR

Over the weekend, Dan and Yosem conferred. At a board meeting on Monday, Dan, Ilya, and Rafi would formally remove Max and decide about appointing Yosem. The battle plan, legally, had been laid out. On Monday, Dan sent a quick note to Yosem: "Meeting has been postponed." Later, in a phone conversation, Dan explained why: they were busy.

Early the next morning, an e-mail arrived from Yosem, sent to Dan, Ilya, and Rafi, but not Max. The subject was "The Story of Diaspora to Date."

"I was not concerned when I heard from Dan that the meeting with Max was being postponed, but I was concerned when I heard about the reason why," Yosem wrote. "Dan said that Dan, Ilya, and Max had had a very productive product brainstorming session and did not want to stop it to talk about the officer role reassignment.

"That's fine, but you need to realize what that decision signals in terms of priorities: 'Since things were going so well at building a product, we could not take time away from that to focus on building a business.'"

The story he laid out, in six thousand words, was primarily his indictment of Max: failed pitches, missed opportunities, the unilateral decision on hiring Sarah. It was tough and true, but not true enough. All four of the Diaspora guys were shareholders in that history, due to either their inattention, indifference, disinclination to mind the nitty-gritty details of being a CEO, failure to have the slightest idea of what they were supposed to do as board members, or, if Max was correct, their mutual decision to turn down the financial support of Kleiner Perkins. (Max's version of the history on that point was not shared by the other three.)

But their ages ran from nineteen to twenty-two. And that youthfulness was not only tied to the mistakes they had made collectively or individually, it was also their treasure. They were the ones who heard Eben Moglen's challenge, who asked themselves, since we do need something better, why shouldn't we build it? They had inspired six thousand people to send them money, and more than one hundred software engineers to contribute code, and Yosem, with his Rolodex of powerful people, to devote the better part of eighteen months to working for them for free.

"What has been accomplished with these resources over nearly two years?" Yosem asked. "Is there a business? No. Have we secured new financing? No. Have we built a great product? Absolutely. But as they say, cool products don't pay the bills."

Diaspora had not found a way to sustain itself. To do so, he argued, it needed leadership that the four of them could not provide, a CEO who had a clue about the business world. "I have recommended that you either appoint me to the position, or if you prefer, I will leave, and you can get someone else with the experience to fill it," Yosem wrote.

The stakes were high.

"Think of the hundreds of thousands of people who have placed all their hopes and dreams in your hands to change the social web for them, and how they would feel if our wrong decisions were to determine their fates adversely," he said. "This is not an issue to take lightly."

It would be Tuesday when they woke to Yosem's message. They realized there would be no meeting on the leadership that day, either. Peter Schurman was getting another fund-raising e-mail message out the door to the public.

"If we are able to get $40,000 per month from this message, as Peter and I hope will happen, we will at least buy some extra time to build the business. But time is running short. We have roughly $20,000 in the bank," Yosem wrote to them.

With the leadership sword fight flashing, Schurman was preparing their largest pitch. For the first time, he was not subjected to the herd-editing techniques that attended his first two informational e-mails, which had not explicitly sought money. Nor had they generated any.

Now Schurman spotted an opportunity in their slow-poke approach to growing the number of users.

Nearly a year earlier, when the founders had launched their private alpha version, all the original contributors had been sent invitations to JoinDiaspora.org, or were supposed to have been, but that was fewer than 7,000. Another 500,000 people who expressed interest, even without contributing, were put on the waiting list.

Of course, it was possible for anyone to download the software and set up a pod on any server, or to join others who were being set up around the world, especially in privacy-conscious pockets of Europe and Latin America. It was no surprise that most people had not attempted to run their own; it was complicated for the nongeekish, who didn't mind having someone else handle their data, just as long as it wasn't Facebook. So they awaited their invitation from the JoinDiaspora mothership.

To Schurman, this was an obvious group to tap: they were interested but did not have an invite. He had inquired casually around the office if he could get some invites. Yeah, yeah, sure, sure, he was told. No one paid any attention or asked why. Besides, the guys were passing out invites all the time, and Yosem was especially prolific at using them to placate irritated members of the public—digital grease.

Schurman's fund-raising pitch was a classic—filled with flattery, coaxing, and promises.

It began: "Dear Friend of Diaspora*—

"We love you. Yes. Really, we do.

"We're building Diaspora*, in a spirit of community, because we believe in you. You're one of the innovators, the creative ones, the people who make the world awesome.

"We're building tools that we hope will help you bring your true voice to the world."

Then the letter turned to the delays in expanding the site. The letter thanked the readers for being "incredibly patient" while waiting for an invitation, and said that Diaspora needed money to build faster.

And then, the payoff: "Also, if you can give any amount at all, we'll be sure to get you an invitation to join us at joindiaspora.com right away. (Just to be clear, you'll still get your invitation regardless. But if you make a gift, we'll get it to you now, so you won't have to wait any longer.)"

As with every public move they made, the headlines were quick, and in some cases, merciless: DIASPORA GETS DESPERATE, ASKS FOR FURTHER COMMUNITY DONATIONS.

One tech blogger said it was "totally absurd" that the guys had paid themselves what amounted to twenty-eight thousand dollars in salary and housing allowances for the previous year. Others thought it was entirely reasonable in a place like San Francisco.

But asking for the money wasn't the problem.

The problem was the promise of an instant entrée for people who gave money. ("We'll be sure to get you an invitation to join us . . . right away.") As it happened, some of the people on the waiting list had already contributed, but somehow never got an invite. Now it looked like they were being hit up again for money. The hooting and hissing grew online.

A few minutes before noon Wednesday, Dan leaned over to Yosem, who was sitting alongside him in the Pivotal offices. Across the table from Dan was Ilya. At another table, sitting with his headphones on, was Max.

"It's happening," Dan said.

Yosem stood and walked over to a coffee table near the Ping-Pong court, followed by Dan, Ilya, and Max.

They sat.

"The three of us have talked," Dan said. "We've agreed to make you the CEO of Diaspora."

His compensation, as Yosem understood it, was to be 3 percent equity, and, once they had gotten financing, compensation equivalent to what the three guys were getting. Dan's manner, usually so casual and laid-back, was strictly business.

"I accept your offer," Yosem said. He noted that he would serve at the pleasure of the board, and that given the uncertain state of his health, he was not likely to be the CEO for a long time. After they had secured financing, they could hire a CEO who had led a series of start-ups—what was known in Silicon Valley as a serial entrepreneur, a person identified less for the products created than for holding a leadership position in companies that were launching.

Ilya seemed delighted that they had turned to Yosem and that the conflict had eased.

Then Max spoke.

"We all talked about it and voted for you," he said. "We wouldn't have let just anyone do it. We feel like you're one of us."

Later, in a phone conversation, Dan told Yosem that he had been surprised Max had gone along with the change.

Fed up with complaints about his communication skills—Max felt the others simply hadn't been listening when he told them things—Max joined the vote. "I'm a team player, whatever you guys decide," he'd said. But he did have a caveat: Yosem should only be an interim CEO, until they could hire someone with heftier credentials.

In the end, Dan said, Max seemed relieved.

"He told me he didn't like being CEO anyway, and this would let him focus on product development," Dan said.

Afterward, when Max offered to get in touch with their start-up lawyer to make the changes, Dan was glad to let him handle the details. He had not enjoyed being so hard on his friend.

First thing Thursday morning, Max checked on their PayPal account, where donors would send contributions. For a moment, all seemed right with the world. Peter's first direct solicitation had gone out on Wednesday, and by eight A.M. the next day, they had received twenty-eight thousand dollars. Plus, it seemed that the pace of giving had gone up overnight.

"It's working!" Max announced in his early morning e-mail to the group.

"Awesome news!:)" Yosem replied.

Casey Grippi, handling their finances, was glad to hear it, too. They needed to put some money into their bank account to pay bills. He logged in to the PayPal account and transferred the bulk of the donations, something close to thirty thousand dollars, to a bank account that had been the repository of their original Kickstarter donations.

But then, some would-be donors sent notes to Yosem, saying that PayPal was rejecting contributions. Max assured the group that the money was still flowing in, though they had gotten inquiries about a few of the transactions, which PayPal seemed to think were fraudulent. At least the spray of donations had managed to interrupt the conversation about who was in charge. For a few hours, anyway.

The next morning, which was Friday, Yosem sent an urgent e-mail. The wording in Peter's message had made it look like they were selling invitations. "We're now in a public relations crisis. Over night, articles and blog entries have appeared all around the world criticizing Diaspora for the perceived 'pay to play' situation. We're also now getting emails from the reporters asking to see whether we have any comments about the negative backlash."

In theory, at least, Yosem had been designated the CEO on Wednesday. By Friday, it was evident that he would not be able to carry out the duties without strife. The question of apologizing became a proxy in the struggle between Max, who was not so quick to cede authority, and Yosem, who wanted to quickly end the invitation controversy.

"I don't agree with this and I don't think we should have such a knee-jerk reaction," Max said. Once everyone got an invitation, things would be okay.

Yosem was insistent.

"Let's not make the classic politician mistake of remaining silent or not saying anything until it becomes a terrible news story," he argued. "We don't have the luxury of losing our 500,000 wait-list members and alienating their friends and supporters."

He listed a series of actions that they needed to take—blogs, e-mails, invitations. "As President and CEO, I will inform Peter that I've decided that his role will remain the same, but that from now on, we will write our own messages," he wrote.

He said a board decision would not be needed to make the apology, but invited them to vote yay or nay on his propositions. "I strongly feel this could break Diaspora as a startup," he said.

Then he drafted an apologetic post, explaining that they had not intended any such favoritism for people with money. He also wrote a Q&A section.

"I'm poor, and I can't donate any money. What can I do to help Diaspora?"

"That's okay, we're poor, too," was the draft answer. It noted that they had started the project as students, that Ilya had dropped out to work on the project, and that Casey Grippi and Yosem were getting no salary.

Ilya didn't like the language. "It's clearly not the case (and doesn't really answer the question)," he wrote late that night. "We are lucky enough to be currently from upper middle class families and to be able to afford to attend an expensive institution such as NYU and have our families' support."

Max wrote the digital shorthand for approval—"+1 Ilya"—and added, "To call a rose a rose, none of the founders are upper middle class. To put it bluntly, we all come from 'the 1%.' (That doesn't mean we are not sympathetic and aware.)"

A version of the apology, with thanks, was worked out. Yosem posted to the Diaspora blog, and e-mailed to people who had complained. He got dozens of e-mails from people who had been glad to get his explanation.

# CHAPTER TWENTY-FIVE

Ilya reported Saturday morning that the money was continuing to roll in, and that the PayPal account now held $41,513.28.

"The issue is well on its way to being controlled, and hopefully, fixed," Yosem wrote.

Max, who had been opposed to the measures taken by Yosem, was openly skeptical. Citing analytics that tracked how often a page was clicked, Max wrote that fewer than two hundred people had viewed the blog post. Still, the others in the conversation were glad to put a few points on the board, and Yosem noted that there had been genuine dismay among their fans at the fund-raising pitch.

"This person captured it most succinctly: 'Every time you guys send me an email, my heart starts jumping, hoping that it will finally be my invitation. In your last message, I excitedly opened it to finally see Diaspora. Instead, you wrote asking for more money.'"

Beyond the squabbling, the coding work had been energized by the need to spiff things up before the launch deadline. They started adding cosmetic features that were reasonably simple changes but made it friendlier: When people created a new profile, they'd be asked "What are you into?" with a series of hashtag prompts for areas of interest. A default first post would be available, introducing new users, a way for others to flag them down and issue a welcome. They'd also find a link to invite others to join the network.

After months of spinning their wheels, Dan thought, the deadline was making them think how Diaspora would really work for people encountering it for the first time. He was charged up.

One new feature took on the major hurdle for Diaspora—the lack of familiar names or friends. At Peter Schurman's suggestion, they added a news stream that carried items not from specific friends or acquaintances—which is how things were set up on Facebook—but items keyed to shared interests. It was nicknamed "soup," and everyone who had tried it, Dan reported, loved it

"Peter—major props," Dan wrote. "I personally cannot wait to invite all these people. They're going to be welcomed with some really kick-ass stuff."

Yosem agreed.

"The next month or so will be a heck of a ride," he wrote.

Positive as these exchanges were, there could be no mistaking that Max was in one camp, Dan and Yosem in another. Ilya was hoping to make his way back to no-man's-land. For the most part, he stayed out of the e-mail fray. He hated conflict and preferred ambassadorial functions, shuttling between feuding parties, and trying to keep from choosing sides. There was no easy way around this one.

In any case, that Saturday night, he would be observing his twenty-second birthday. The theme of the party, he had announced in an e-mail, was "FuckYeahCarlSegen," misspelling the astronomer's name. Dan tweeted the news that he would be the DJ:

busting out the gear. dj set tonight @ Illya Zhitomirskiys #electrofunhouse

Elizabeth Stark arrived at the Hive with a piece of key lime pie and a candle as the birthday cake proxy, and also in honor of a memorable outing they'd had a few months earlier. She, Ilya, and another friend had gone to Cafe Gratitude in the Mission, an über-Californian restaurant devoted to raw and vegan foods. They all had key lime pie and Ilya liked it so much that he could not get it out of his mind. Much later that night, after they'd left the restaurant, the taste of it lingered. He persuaded Stark and the other friend to go back to the café. It was closing for the night. Done. Ilya had to have a piece of key lime pie. He cajoled a waitress who was still there, until, finally, she said: "I'll make you a deal you can't refuse. You give me ten dollars, and I'll give you the whole pie."

Ilya had bragged for days about this coup, and Stark knew how happy he'd be to have a piece on his birthday.

For many of the guests, what was most memorable about the evening was not its theme but Ilya's condition. Yosem was shocked by his level of intoxication.

A few days earlier, an NYU friend, Aditi Rajaram, had texted him for his birthday. She had recently moved to California to work for Google. He immediately, enthusiastically, asked her to join the party. At school, Aditi had been inseparable from Max, Rafi, Dan, and Ilya. When their Kickstarter campaign had launched, she was one of the people who got automatic notifications; she also had the password to the group's Dropbox, a cloud storage account. Their departure had been hard for her.

She felt particularly close to Max, and had sat next to him at his graduation party in the Tribeca loft, watching the tide of donations roll across the screen. Once they moved to San Francisco, she found herself cut off from him, as she was still at NYU, with senior year ahead. To the regret of both, he was often too absorbed with work to answer her e-mails or texts. He and Ilya had been her informal tutors on matters of technology. Their patience had been a gift; she was a liberal arts major. Ilya had a knack for cracking open the shells around what seemed like hard concepts.

One day he found her leaning over books, trying to get her mind around the concepts of encryption. Ilya explained to her that it worked like a lock and key, with the keys handed only to those allowed in.

After the movie *Inception* came out, Ilya spoke with her about lucid dreaming—a belief that people can, at some conscious level, exercise control of their dreams, permitting themselves, for instance, to fly. During a romantic interlude in their friendship, they hatched a plan to do the lucid dreaming together, but never got around to it. She missed all of them. When Ilya had fallen into depression a year earlier, around Thanksgiving 2010, she was startled by the darkness of his emotional hues. They stayed in touch by phone.

That fall, she had moved to San Francisco, but her new job and their long hours formed a moat between them. Over time, she, too, had heard accounts from each of the four about the dynamics of the project. Life had gotten infinitely more complicated than it had been back on Mercer Street in Greenwich Village. Each one had stories so different that it brought to mind for her *The Great Gatsby,* which was often taught as a prime example of an unreliable narrator. She thought Rafi had the clearest, most grounded vision of what was going on. Still, she was very happy to be invited to Ilya's party; to reconnect with people who had meant so much to her.

At the Hive, Ilya greeted her exuberantly, but there was something off-kilter; the emotional whitecaps of his language and manner were almost too choppy for her to navigate.

From across the backyard, another guest, Katie Johnson, spotted Ilya and had the same reaction as Aditi. He was not taking care of himself, she thought.

Katie was a member of one of Ilya's early crops of friends in San Francisco, the people he had met in early 2011 after returning from the breakdown. She was the one who had invited him to the brunch back in June, only to see him tumble off his bike as he followed a pickup truck, trying to ride while simultaneously sipping a Bloody Mary.

Over the summer, they got together regularly to talk through their projects. It was a sociable, but not social, relationship.

Katie had been working with an educational start-up. Ilya had an idea for creating a party app, Epic Parties, that would make it easier to organize parties and, for the people who met one another, to stay in touch later through social media.

He ordered long rolls of white paper so that any ideas or plans they sketched out would not have to fit on an 8½ x 11 inch sheet of paper. They created a tongue-in-cheek scroll for a book called *Stuff That Leaders Do,* with items like, the leader makes sure everyone is included before walking anywhere (when a group would mosey along in tattered formation to the park), or, when the beer ran out, the leader would say, "I'll go get it."

Most of the time, their conversations were about work. One time, they'd gone for a falafel and talked about relationships. He was concerned that a younger girlfriend had become too attached to him, and Katie suggested some gentle ways to detach. She looked forward, always, to their get-togethers: they were ritual renewals, cheerleading sessions. One day in the summer, she got an e-mail from him: Hey, I really want to be your friend, regardless, but if you would ever want to be in a relationship with me, I would really like that. You're awesome, he had written, no matter what.

It was a note full of care and risk. Just what you would hope the modern male would be capable of, Katie thought. He was so intuitive and empathetic; he paid attention and was full of energy. She wrote back, telling him that she knew it had not been easy for him to let her know that this was on his mind. It took so much. And it was so great. "But I really see us being friends," she said.

Their friendship never missed a beat. But ever since the end of summer at Burning Man, where she had watched him racing about the playa, she felt slightly unsettled by him. He seemed to be working all hours of the day. Yet he remained endlessly, insistently generous. She had a crisis in September: the start-up she was working for had relocated to New York. Ilya was the one person she wanted to talk to about what she could possibly do next, and he had gone with her to brunch at the BrainWash Cafe, a combination Laundromat and dining spot. He made her describe her interests—environmentalism, outdoor activities, bringing people together, collaborative consumption—and think of where they intersected. And that was it: she would create an app for people to organize walks with others. She would call it "hikery."

It was a fine session, and Katie was grateful that he had been able to make time for her. He also had started calling various friends to talk up creating nighttime sports leagues, a healthy alternative to the bar life. It was a way to save the world, he said. To Katie, his own health did not seem to be in such a prime state, but maybe that was why he was thinking of alternatives.

A week before Ilya's birthday, the Burning Man festival was holding its annual "Decompression" session at a big parking lot in Potrero Hill. It was a one-day reprise of the festival, with performances and art exhibits and reconnections. Ilya's bedroom window looked toward Potrero Hill, a twenty-minute walk from the house.

Katie had called him.

"Are you going to Decompression?" she asked.

"I really don't have time," Ilya said. "I'd love to get together. I'm going for pizza now."

They arranged a quick visit, and ended up sitting on the porch at the Hive, talking about the pressures at Diaspora. He made no mention of the burgeoning conflicts over leadership, but told her that he was concerned about the people who had contributed to the project, and not letting them down. He was stoked, though, that they had hired Peter Schurman to organize their mailings. Then she went on to the Decompression event and he went back upstairs to his laptop.

Now, at the party a week later, he threw his arms around her in welcome. He was very, very high. It seemed like he had not showered in a while, and he seemed exhausted. "Don't work too hard," Katie said.

Before the night was over, a woman Ilya had recently started seeing gave him Special K, ketamine. Its pharmaceutical origins were as an anesthetic used in veterinary medicine. On this occasion, it was meant as a celebratory treat: a drug with short-lived hallucinatory properties typically used in club scenes.

# CHAPTER TWENTY-SIX

The following week, all work getting ready for the beta launch of Diaspora came to a halt. PayPal had frozen its account—something had tripped an internal monitoring mechanism, and the money company apparently became concerned that Diaspora was some kind of flimflam being run through the PayPal account. It took back the money already transferred to their bank. A review would take 180 days.

Diaspora mounted a digital counterinsurgency. Peter wrote a press release and Yosem gave interviews to the tech press. Dan, Max, and Ilya tweeted to the sixty thousand Diaspora followers on Twitter, then hundreds of community followers also posted tweets, calling on PayPal to unfreeze the Diaspora account. Then WikiLeaks, the digital vault created by Julian Assange to receive and transmit secret documents, weighed in from its Twitter account. They, too, demanded that the Diaspora account be unfrozen.

WikiLeaks had famously been blockaded from a number of commercial services, including its PayPal accounts in November 2010, when it began to roll out a cache of documents that had been given to it by a private in the U.S. Army. They included military reports and videos of the shooting of civilians in Iraq, and thousands of diplomatic cables that lacked the usual coats of varnish about how the United States viewed various parties it was dealing with around the world.

WikiLeaks was followed by more than 1 million people on Twitter, so it was a powerful amplifier for Diaspora's campaign. After two days, PayPal decided it did not need six months to figure out what was going on. It unfroze the accounts. For all the useless angst the episode had caused, they'd generated plenty of useful sympathy.

Once the smoke from the PayPal crisis had lifted, Yosem and Casey began pricing office space. It was time for them to leave the nurturing womb at Pivotal.

In less than a month, mid-November, they would be launching. Volunteers and the Diaspora community, long neglected, were being stroked by Yosem, invited to take on roles in welcoming new users—a touch that would distinguish Diaspora from Facebook and Google. As the tasks became more concrete, Yosem wanted to be able to tell them that he was not only functioning as the CEO, he *was* the CEO.

"Where do things stand with the legal paperwork?" Yosem asked Dan.

Actually, nothing had happened at all.

"I asked Max to handle it," Dan said. "I felt bad for him. I wanted him to feel like he was still part of things."

At a human level, Dan's response made perfect sense. He cared about Max. As a business question, it defied human nature.

"Duh," Yosem said. "Basically, you fire him as CEO and then you tell him to go through the process of filling out the paperwork for the new CEO. I'm sure he's not excited to do this task."

"Why don't you do it?" Dan suggested.

For help getting it done, Yosem turned to Diaspora's lawyers and to Casey. They got the paperwork together for a board meeting.

They convened on Monday. Julian Assange, the founder of WikiLeaks then on the run from authorities in Switzerland, had mentioned the PayPal blockade of Diaspora in a press conference that afternoon. "Badass!" Dan exclaimed.

The hunt for space would continue that week, with Casey and Yosem carrying the ball. Management of the community was a beast, but there were plans now for online meetings in Internet Relay Chat rooms. The night before, Ilya had called Evan Korth, their NYU mentor, to break the news that they were reorganizing the board, and that they would like him to step down. The board had originally been the four guys, with Korth added solely as a tiebreaker in the event of a deadlock. With Rafi about to step down, that role would no longer be needed. Still, it was a sensitive conversation, and Ilya knew that even though Korth was a completely good sport about it, he was a touch hurt.

"We need to send him a letter that he can sign and formally resign," Ilya reported.

Over the weekend, Dan, Ilya, Casey, and the lawyers had gone through the details about Yosem's appointment. Max had not been included in those exchanges, but he had gotten them that morning, ahead of the meeting.

"The paperwork is going into the lawyers, and Casey is going to handle the changes to the board and naming me as CEO in the papers," Yosem said.

Ilya spoke up.

"Oh," he said. "We have to revisit that, to be sure that we approved everything."

This was a startling development: everything had been shared well in advance of the meeting. Dan and Yosem were perplexed.

"What is there left to approve?" Yosem asked. "You already offered me this position. I thought that the terms were set."

Some other details had to be discussed, Ilya told him. After the meeting, Dan approached Yosem and said he had been surprised that Ilya had put on the brakes.

"I had no idea," he said, promising to find out what was going on.

In fact, Max and Ilya had discussed Yosem's appointment as CEO earlier in the day. Max insisted that he had gone along with Yosem's appointment because everyone agreed that he would serve only as interim CEO while they searched for an established entrepreneur. Moreover, his equity compensation should not vest until he had been there six months. That was not what the terms drafted by the lawyers said: Yosem served at the pleasure of the board, effectively an interim appointment, but that if he were terminated, his shares would vest immediately.

Dan huddled with Ilya after the meeting, then called Yosem. He did not get into the nuts and bolts of the compensation.

"I guess we never actually made a motion about it," Dan said. "We just talked about it with Max and we decided to give you this, but I guess Ilya says we have to pass a motion on it."

That meant a formal board meeting. They would get to it the next day, Tuesday.

"That's fine," Yosem said.

Home in Palo Alto that night, he spoke to his fiancée, Emily, about the developments. "Before they offered me this role of CEO, I used to enjoy the business," Yosem said. "I was doing it because it was a cause I believed in. Now I'm CEO and I absolutely hate what I'm doing."

On Tuesday morning, they could not have a board meeting: Ilya was due to speak on a panel at the Silicon Valley Human Rights Conference being held that day in San Francisco. He had been in a lather for days about what he would say. The subject: "Social Networks: Exploring How Defending Human Rights Can Actually Strengthen a Company's Bottom Line." He was quite pleased to have been invited, though he knew that Diaspora did not exactly have a bottom line. But he got through it.

Afterward, T. H. Nguyen, a young lawyer, approached Ilya. Having recently left Facebook, she was searching for work and a purpose. Already she had raced through the online material about Diaspora. It was clear that they were neophytes who had little experience figuring out how to make their users happy. She could help.

"That would be awesome," Ilya said. "We'd love to have you. Please, get in touch."

After the human rights conference had formally broken for the day, the sponsors threw a drinks party at a bar in the Mission. Elizabeth Stark had gone, and Ilya was with her. Samuel J. Klein, known as SJ, was visiting from the East Coast and had been crashing with Stark. But that wasn't the main reason for his coming to the party. He wanted to meet Ilya.

They were both committed members of the tribe of technologist-idealists. SJ, a Harvard-educated physicist, had been among the earliest contributors to Wikipedia, the crowdsourced encyclopedia, starting in 2004, when, as he explained it, the site had "officially become more interesting and instructing than universities and course books." He was a manager with the One Laptop Per Child project started by Nicholas

Negroponte, the cofounder of the media lab at MIT. Its goal was to build network-able laptops for one hundred dollars or less and distribute them to children in the developing world. The project had been criticized for ways in which it fell short of its soaring ambition.

When Diaspora was getting started, SJ had written to them, suggesting that they get a secretary to manage their correspondence and make sure that useful ideas did not disappear behind the cushions of sofas.

At the bar, he mentioned that letter to Ilya, who laughed and said it would have been a great idea.

Even at conferences like this, the social encounters could be overwhelming. They got talking about how to manage that. Both had been toying with ideas for an app that would capture promising contacts from such fleeting connections, but also make them part of a social media structure.

As they spoke, an attractive woman came up, smiling. She spoke to Ilya:

"You made my day," she said.

"I did?" Ilya said, laughing, embarrassed.

She had come to the conference, and he had been a singularly inspiring voice. So positive. They fell into conversation; they spoke long enough for Ilya to hear about her plans and wish her well.

Elizabeth, meanwhile, was getting ready to go back to Palo Alto, and SJ, her guest, was plainly not ready to go. She turned to Ilya.

"Can SJ stay in the city with you?" she asked.

"Totally," Ilya said.

Ilya gave him the address on Treat Avenue, and told SJ to let himself in whenever he was through.

A few hours later, SJ arrived at the Hive and found Ilya hacking at ways to create social graphs that connected interesting people he'd met at conferences. Their brainstorms spread down a long sheet of butcher paper.

The first item on the list was their goal for the business cards that were stuffed into their pockets at the end of such conferences: "Get Rid of Effing Cards."

They moved through more items, concluding with a note to integrate "epic projects," another dream child of Ilya's. They looked at the time, and realized it had gotten to be four A.M., and both had things to do in the morning. Ilya snapped a picture of the page for safekeeping and e-mailed it to Klein.

Before turning in, Ilya mentioned one other project: breaking the world record for the longest chain of linked plastic monkeys. They had tried it earlier that summer with Parker Phinney, another drop-in guest at the Hive. Someone had leaned out the window in Ilya's bedroom on the top floor, but there were limits to how many monkeys could be hung from one another. But the constraint seemed to be the weight of so many monkeys: the chain would collapse before they could get to the record.

"It's a physics problem," Klein said.

As they talked it through, they came up with a likely solution: weave a series of loops into the configuration, breaking the straight downward load. Ilya could picture it.

"Awesome," he said.

A few hours later on Wednesday morning, Casey and Yosem were crunching numbers. How much were they paying for every server? How much longer would they be paying expenses? What could they afford in rent if they were able to get new space? In short, how much of their resources were they burning every month, and how long could they last?

Dan called Yosem about the CEO job.

"Is there a way to go ahead with the plans without having to hold a meeting with Max again?" Dan asked.

That made sense to Yosem. "I don't think it is a good idea to rehash this again with Max," he said. "To the extent that Max is getting over it, this is going to put him right back in the position of getting upset again."

They checked and discovered that they could make the changes by e-mail. Max then insisted that he, Ilya, and Dan hold an emergency board meeting.

Yosem headed for the offices of the Wikimedia Foundation, the parent of the online encyclopedia Wikipedia. They were willing to host Diaspora free of charge. This was a pleasant development. As he was heading home, the phone rang. The guys had been meeting for three or four hours, and they were all on a conference call. As Yosem and Dan recalled the conversation, Max did most of the talking.

"Yosem, we have to renegotiate your terms," Max said. "First of all, we never agreed to give you a seat on the board."

"That was part of the agreement," Yosem said.

"If it was, we don't really remember it," Max said. "We need to discuss that again, but it just really doesn't make sense for us to do that."

Yosem would have essentially replaced Evan Korth as the tiebreaker, but since Rafi was going to resign from the board, there would be no need for a tiebreaker. Moreover, a smaller board made sense because investors would want a seat.

"In that case, if you want to try to have fewer people, I don't need to be on the board," Yosem said.

There was another item, apparently the result of a misunderstanding, Max said. His 3 percent share would be vested over four years, not one. Also, they would not be changing the papers of incorporation, which still listed Max as CEO, and substituting Yosem.

"We also don't want to put you down as president officially. We just want to give you the role," Max said. "We just want to have you as an employee. That way, if we find someone down the line who is better than you, we can replace you. You can just continue working for us in another role."

"That's okay," Yosem said more reflexively than reflectively.

It had always been his intention to serve on an interim basis. These elements were worked into the contract that had been prepared and, he thought, agreed upon already with Dan and Ilya. Perhaps that was what Max was telling him.

After they hung up, he replayed the conversation and, as the details settled on him, was stunned. The phone rang again. It was Dan.

"Hey, I really want to apologize for that. We really want you to just choose whatever terms you want. I don't know why Max was saying those things," Dan said.

As he continued speaking, Yosem was unable to absorb much of what Dan was saying.

"Okay," Yosem said. "Thanks."

He drove home.

Near six that day, Mike Sofaer approached Ilya about his plans for the evening. He was aware of the conflict between Max and Yosem, but did not know specifically that the tension was cresting.

Mike had something else on his mind entirely: he wanted to check out the Occupy Oakland encampment, a sprout of the movement calling for economic justice that had started under the banner of Occupy Wall Street. Nearly from the beginning of the Occupy Wall Street protests in the fall of 2010, Ilya had been an avid follower, at least on his computer screen. He had a live stream of the protests running virtually all the time back in the Hive. When the movement spread across the country, the demonstrations and occupations became must-see spectacles for many people. It was not clear now, at the end of October, how much longer they would last, at least in Oakland and San Francisco. Officials in San Francisco were giving signals, widely reported online and in the mainstream press, that police were preparing to raid and roust the San Francisco encampment on the Embarcadero that night. And the day before, the authorities in Oakland had used rubber bullets and batons to clear protesters. Word of the clashes hummed on the social media lines. Mike had been an occasional confidant and advocate of Ilya's for a year, ever since Ilya was wracked with doubts of his worthiness for the project. Ilya became a regular at Mike's occasional board game nights, a gathering of low-key, sociable geekery. A visit to the Occupy camp would, at least, be a change of venue for Ilya's tension. He agreed to go along.

They left the Pivotal offices on Market Street and grabbed the subway for the twenty-minute ride to downtown Oakland. At city hall, they headed to the Frank H. Ogawa Plaza, which the protesters had renamed Oscar Grant Plaza after an unarmed young man who had been shot and killed on a subway platform by a police officer.

When they got there, the protestors were heaving metal barricades, triumphantly dismantling corrals set up by the police. The scene was chaotic, and unnerved Ilya. He and Mike walked across the plaza to a Subway hero shop, bought sandwiches, and watched the events as they ate. Then a call went out for people to gather across the bay in San Francisco, on the Embarcadero. Mike planned to go.

They headed back across the bay to San Francisco on the BART. At the Embarcadero stop, as Mike prepared to get out, Ilya said, "I've got a ton of work." He continued on his way to the Hive.

An e-mail was waiting for him when he got out of the subway.

From: Yosem Companys <yosem@joindiaspora.com>

Date: Wed, Oct 26, 2011 at 8:09 PM

Subject: I quit

To: Daniel Vincent Grippi <daniel@joindiaspora.com>, Ilya Zhitomirskiy <ilya@joindiaspora.com>, Maxwell Salzberg <maxwell@joindiaspora.com>

It pains me a lot to say this, as I love you guys, and I love Diaspora. I believe in the cause. But I'm going to get straight to the point: I quit.

After his phone call with Max, and then Dan's call saying that everything could be worked out, Yosem had spent the afternoon and evening coming to a view that he was stuck in a rerun of the last nine months. The precise pattern he and Dan had discussed at the beginning of October, of decisions made and then countermanded by Max, was made plain in the events of that day. He did not need to think about his decision any longer.

"We've done everything possible over the past few weeks, so Max wouldn't feel bad about the situation, even though he was the one who caused it by at best engaging in poor communication and at worst lying to everyone.

"For a week or two after the situation arose with Max, I mentioned to Dan that Diaspora had for the first time stopped being fun for me. It had been fun the whole time prior to that, but Max just made it impossible for everyone to continue having fun. And yet after the whole situation Max acted as though we had done something to offend him, when in any company under those circumstances he would have been fired, and no one would have cared one iota about his feelings.

"So I've come to the conclusion that there is only one solution: Either I go, or Max goes. Since I know you guys are friends and have known each other longer than you know me, I will just make it easy for you: I go."

Later that night, the phone rang, first Dan, then Ilya, asking him to reconsider. Yosem said his terms were firm.

"I don't think Max has to leave completely, but he definitely has to be off the board. If Max agrees not to be involved in any decision-making function for the business, then I will come back. Otherwise, I won't."

# CHAPTER TWENTY-SEVEN

"**Y**ou need to talk to Max," Ilya said. "If you guys talk on the phone, you will be able to resolve this."

In fact, the situation had become irreconcilable.

Max called Yosem the following day. Max had valued his counsel for more than a year. More than anyone else, the two of them had thought about getting Diaspora to a sustainable place. For the last month, he felt, Yosem simply was cutting him out.

"I'm sorry you're upset about this," Max said.

As he saw it, the others were not remembering things correctly; they simply were inattentive to most aspects of the business, wanting to just code and code.

Yosem said that Ilya, Dan, and Rafi had all been consistent in their versions of what had happened. "Either these three people have really terrible memories, or they are just doing this on purpose because of some kind of conspiracy against you, or you're just not telling the truth," Yosem said. "If you made a mistake, just admit it and say it."

"I've already apologized," Max said. "I've told you that I'm sorry for you being upset."

They briefly discussed the hiring of Sarah, which Yosem had assumed the others knew about. Max said it had always been part of the plan, and that it was in the shared bank of documents that all of them had access to.

"Sarah is a core part of the team and she will come here," Max said.

The conversation ended. Incensed, Yosem emended the "I quit" e-mail that he had sent the previous day. "Just to clarify my conditions," he wrote. Not only was he insisting that Max not be a director: "Sarah is not coming on board, either."

All around, the relationships were rubble. Dan felt himself being twisted. Initially incensed by what he saw as Max's presumption, he was now shocked by Yosem's assertiveness. He wanted Yosem to be the CEO, but now Max was insisting that Yosem was acting crazy.

Dan called Yosem again, said he would come to Palo Alto on Saturday, and that they would work it out. Nevertheless, Yosem started forwarding any important e-mail traffic, like press inquiries, to Dan and Peter, though he continued taking care of small bits of business, like people who wrote in requesting an invitation to the site, or had other minor administrative problems. He sent out a note to reporters on the press list, advising them that the project was going to have a beta launch on November 15.

He got another call from Dan on Friday, the twenty-eighth.

"Got some good news," Dan said, telling him about the discovery of an error in an automated e-mail system. People hadn't stopped joining or started ignoring Diaspora; the e-mail program had accidentally stopped inviting them to join. They would not have a hard time hitting the 2 million mark by the end of 2011.

"That's great," Yosem said.

"Yesterday, we had a really good meeting with Max, and we realized the product is going to do really well when we do the launch, and we're going to have no trouble raising financing after that. And actually we're kind of upset, because we feel that you've left us hanging out to dry. You quit two days ago and now everything is falling apart because you haven't been doing anything."

"I have been working," Yosem said. "I've been depressed and I've been waiting for you guys to resolve the situation, but I've been working and I've been doing stuff."

Dan also pointed out that Yosem had most of the master passwords and account names for Diaspora's business functions. "We don't know any of that stuff," Dan said.

"What are you talking about?" Yosem said. "Max is the one who created all these accounts, and he gave me all that information. I haven't changed any of it."

"Oh," Dan said.

Yosem was flabbergasted. Having worked for more than a year without any compensation or promise of it, he did not feel there was any call for him to be blamed for neglecting the interests of the company. He stayed up all night, wrote a quasi-transition plan, and included a list of all the passwords and user names for the company. He cleaned out his in-box and forwarded everything to Dan and Peter. One interesting piece of e-mail that did not go to them was from an organizer of the Occupy Philadelphia movement. He wanted to know if it was possible to adapt Diaspora as a private network among the activists. Yosem forwarded that to Ilya.

At four on Saturday afternoon, he sent out a short announcement to the entire community, to which only a few weeks earlier he had been introduced as the CEO.

"Due to personal reasons, I've left Diaspora," Yosem wrote. He gave Peter Schurman's e-mail address for inquiries.

On Monday, Halloween, as Yosem and I spoke about the trail of events, I could hear his e-mail alert bell ringing in the background. A message arrived from Dan, and it was addressed to the Diaspora grassroots group. The subject was "About Yosem and What's Next." People had been inquiring about the departure of Yosem.

"We loved working with Yosem and he did really great work while he was here, especially his work with all of you. We're really sad that he's gone.

"His note raised questions, however, and with all the energy all of you give to Diaspora*, you more than deserve an explanation. So here's the deal.

"For a long time, we had discussed with him the possibility of him becoming not just an awesome advisor but the President of the Diaspora* Foundation. A couple of weeks ago, in discussions about what terms would go with that new role, we agreed to consider some things he had proposed. Nothing was formally agreed to or signed, but at that stage he announced to the community that he was President. Last week, it became clear that we couldn't meet the terms that he was asking for. At that point, suddenly, he left.

"It came at a tough time for all of us, so close to our upcoming launch. The launch will be huge, and it's just around the corner, and there's a ton to do. It will be a challenge to pull off with Yosem gone, but we know we can do it with your help."

As Yosem scrolled through Dan's e-mail, Emily came home from her shift as a hospital resident, and read over his shoulder.

He tried to figure out who had written the letter, saying that much of the language appeared to be the work of his friend Peter—apart from the direct and, he said, completely disingenuous description of what had gone on with him. Nevertheless, it had gone out in Dan's e-mail.

"I love Dan—Dan's my favorite guy," Yosem said to me on the phone.

In the background, Emily chimed in: "Not anymore."

"Dan was always my favorite among them," Yosem said. "He was very honest when you would talk to him one-to-one."

"Until today," Emily said.

The day after we spoke, he sent a reply to the same list that Dan had addressed, saying he had not intended to make any further comment, but did want to post one last thought.

"I was pretty upset when I read their account, a little bit because what they wrote is a blatant lie, but mostly because their description makes me sound like some power-grabbing impulsive whack job, and I may be many things but I'm not greedy, I'm not power hungry, and I really don't think I'm an impulsive whack job.

"The truth of the matter is that I have worked for Diaspora* for the past year-and-a-half without ever receiving a single dollar in compensation. I wasn't doing it for the money—I was doing it solely because I believe in the idea, and it was fun for me for a very long time. And then, it wasn't fun any more. I learned relatively recently that the business was not being managed the way I thought it should have been.

"So I wrote to the board and said that either they needed to find new leadership—me or somebody else—or I wouldn't be able to continue to be involved. So they offered me the position of President and CEO (still unpaid, of course, since we don't have a lot of extra money), and I accepted. And I did write some messages referring to myself as President and CEO because I had been told I was. And then I found out that, although the offer had been made, no legal changes had occurred to that effect. When I insisted that they change my legal status, the board told me they were reconsidering the offer and wanted to negotiate some more. So at that point, I put my foot down and said things either had to change immediately, or I was going to resign as President and CEO and leave Diaspora entirely. As you all know, they decided to let me go. And so I went."

This would be his last message on the subject, Yosem promised.

"I will also try to remember what it felt like when I was 23-years-old and somebody criticized or ditched me, and my feelings got hurt. I'm sure I probably also reacted in a way that was less than impressive. And I'm sure 10 years from now when they are my age they will look back on their behavior and be a little bit embarrassed or ashamed of it all. But then again, isn't embarrassing yourself the entire point of your early 20s anyway? :)"

The message was sent out just before one-thirty A.M. on November 1.

A few minutes later, Ilya called Yosem, who didn't hear the phone ring. He only noticed the missed call a few days later. Neither of them tried again. What, after all, would be the point?

The departure of Yosem galvanized the remaining three members of the team. Peter was still working with them. Casey, who had tried to broker a peace, stayed on, wondering how they were going to pick up the slack. Dan's anger at Max had shifted, slightly, to Yosem: how could he walk on them at this moment? Dan, always rigorous about punctuality— he wanted a clean start to the day, and a clean end—insisted that they meet every day. He turned his cross-country running endurance skills to coding.

On the Sunday after Yosem quit, Dan tweeted:

is it sadistic of me to say that stress is the best driver for just about anything? #crazysunday #gogogo #keepcalmandcarryon

A moment later, he realized he was not talking about inflicting pain on others, and corrected himself, invoking Homer Simpson.

*Doh masochist

They huddled at the Pivotal tables in hoodies, pasty-faced, racing to meet their November 15 deadline. A big part of that was managing the community and marshaling it to welcome new users and help them find interests and interesting people. Ilya, who had jurisdiction over that part of the operation, was not keeping up with the flood of inquiries about the departure of Yosem and the anger of the devoted early users. It seemed to Max unlikely that they would make the beta release deadline, but the exercise was worth it.

Dan tweeted a picture of a glassy-eyed Max, elbows on the table, over his laptop; in the background, Ilya was gazing into his machine. He captioned it:

so many #nerdproblems right now

In the midst of their sprint, they got news that delighted them.

A Diaspora user in France, David Ammouial, posted a personal announcement.

"I'm a little embarrassed to announce that Diaspora has become a dating service, too. Yeah, don't laugh. Sometime ago, I started to follow somebody's posts (I won't disclose any name), then they started to follow mine as well, and we slowly got to know and appreciate each other's ideas and interests. Since Diaspora doesn't have a chat (yet), we began to chat via Jabber/GTalk and progressively became closer and like each other more. We finally met in real life recently . . . and are now in a relationship. I didn't think using Diaspora could lead to that.

Ilya spotted the post and commented:

OMG This is amazing!!! ♥

Dan weighed in, too:

THIS IS AWESOME. #CONGRATS!

It was, Dan said, soulful news: people could meet one another on Diaspora; they could discover love there. What more was needed? Social networks fancied themselves a more evolved form of dating websites, which were limited-purpose auction houses. A much broader spectrum of human connection was possible on the social web. But other networks were not set up for this kind of chance encounter between two strangers. On Diaspora, committees of existing members welcomed new users, almost concierges, precisely to help people feel not so isolated in a network that was so devoid of existing friends and acquaintances. That serendipity was part of their fund-raising pitch, and David Ammouial's happy news was a perfect example.

The next afternoon, Dan was using automated programs to see how Diaspora's code held up under simulated use. The load on his computer was immense. He tweeted:

using my macbook pro as a second heater by running #diaspora's test suite

# CHAPTER TWENTY-EIGHT

**B**obby Fishkin never aimed low during meetings of his association—
after all, it was named Visionaries and Revolutionaries—but his
question at the November meeting stumped the audience. It was
the first Sunday of November, the sixth, and they had gathered in a room
at the Yerba Buena Center for the Arts. That evening's topic was inspired
by Gottfried Wilhelm Leibniz, a mathematician and philosopher of the
seventeenth century. The name was not widely known, but his work and
life had intrigued Bobby for fifteen years, so he gave a short preamble
before throwing out his question.

"The claim is that Leibniz was the last person in human history to
know everything about all disciplines," Bobby said. "That he learned
everything that had been declared to be knowledge up until that point in
human history."

Today, he said, individuals were silo'd into their fields and social networks didn't cross disciplinary lines. "There are ten thousand subdisci-
plines," he said. "Could the metallurgy of diplomacy be of value? Could
the physics of sociology be meaningful? Could the epidemiology of history be an important area for study?"

Some of these combinations, perhaps many, were of no importance.
But, he said, "A subset of all those hypothetical disciplines are of profound
relevance to human challenges. Yet with all the trillion hypothetical

convergences of disciplines, we're never going to get around to enough of them."

Here, he said, was a Leibniz challenge: how to encourage enough cross-pollination of disciplines to ensure that useful hybrids were created.

"What would be the most efficient ways of creating the greatest diversity of solutions with the greatest possible relevance to solving human challenges?" he asked the group.

A few times, Bobby scanned the room to see if Ilya had arrived; he had said he was coming, but never showed. Bobby was annoyed. But Tony Lai, the lawyer roommate in the Hive, had come.

The conversations about the challenge were stimulating, but by the end of the evening, no one had come up with anything approaching a satisfactory answer. The center closed at eight P.M. on Sundays. Afterward, Tony invited Bobby back to the Hive.

Ilya was in his bedroom, working at his computer, and at first did not come out to visit. He loaded an image of a giant butterfly, ascending to the heavens from a tropical garden, onto his Diaspora feed. After some cajoling, he joined them in the living room. Bobby told him about the Leibniz challenge.

"Everyone was stumped," Fishkin said.

"Really?" Ilya asked. He gave it a moment then said that the answer was straightforward: make it a requirement for high school graduation that every student work on a cross-disciplinary project aimed at improving the world.

"In twenty years, you would have a critical mass of collaborations," Ilya said. "You'd have a statistically significant number of people who had crossed disciplinary lines."

In the span of a generation, he said, you'd have enough people to take on epic challenges.

The night rolled on, the conversations sprawling across acres. Ilya had been blown away by an interview with Jaron Lanier, a computer scientist, pioneer of virtual reality, composer, and author. Lanier described a "local-global flip" in which an illusion of power was gained through networks, with a "heroin-like" allure. People needed to understand that there was no such thing as "free," Lanier argued. Uploading a personal video to YouTube, he said, was working to create fortunes for someone else.

"If you're adding to the network, do you expect anything back from it? And since we've been hypnotized in the last eleven or twelve years into thinking that we shouldn't expect anything for what we do with our hearts or our minds online, we think that our own contributions aren't worth money."

Lanier said Apple was funneling information through its hub, asserting control over how people consumed art that they owned. Google had decided that "'since Moore's law makes computation really cheap, let's just give away the computation, but keep the data.' And that's a disaster."

Bobby said he looked forward to reading the entire Lanier interview, and Ilya sent him the link.

They talked about their projects. Besides Diaspora, Ilya had gone back to his plans to create the Epic Parties app, and was thinking of ways to open spontaneous hacker spaces. The idea was that he would get half of the unaffiliated hackers to sign up for one, kicking in some modest membership fee, and use some of the revenue to open yet another, and so on. It had elements of genius and Ponzi at the same time.

More pointed, Ilya had an idea to pinch off the illegal-drug trade at the bottom of the chain. He had read a study that said most workers in illegal-drug organizations had very little money and were not enjoying romantic liaisons.

Billboards, he believed, could take out drug dealers by simply making it embarrassing to be one. He had sample messages in mind:

"Drug Dealers on Average Make Less Than the Minimum Wage.

"They Live with Their Mommas.

"They Don't Get Laid." It was a low-cost and safe tactic, Ilya said, noting, "Some of the dragons are paper tigers."

Diaspora was consuming enormous amounts of time. They were hoping to get the beta release out, but that was looking less and less likely with Yosem gone. Ilya could have stayed up later talking, but Fishkin had a flight to catch in the morning, so after three hours, he took his leave.

"We have too many revolutions, Bobby," Ilya said. "This is good. We've got to hang out more."

———

Dan had spent the evening watching an episode of *Breaking Bad,* the neogothic series about a nerdy science teacher who becomes the master meth kingpin of the southwestern United States.

He tweeted:

albuquerque, NM is a fictional place where not a single person has graduated from the flip-phone. #Breakingbad

Addressing the main character, he noted:

walter white, you've gone on for 5 seasons without checking twitter once. #Breakingbad

Ilya might have been barely treading water, but Dan was exhilarated by the demands, and he let the world know in a series of tweets.

we successfully migrated jd's database to a 16gb o' ram box. #diaspora #scaling #phew

having the best support convo with alison o. from @rackspace right now. #Happypanda

@newrelic you guys are fucking legit. thank you SO MUCH for server monitoring. ♥

we're back up! major props to Sarah Mei Maxwell Salzberg Ilya Zhitomirskiy #♥ #diaspora #scaling #scalingmeansmoreusersarecoming

nyc in 3 days. NEW YORK, I NEED YOU.

Dan was going to speak at NYU that weekend on the delicate circumstances of people starting up tech ventures with friends and classmates, just as Mark Zuckerberg had done with Facebook. The talk was called "Our Zucks."

The next day, Monday, November 7, the technology news section of the *Wall Street Journal* ran a blog item, "Whatever Happened to Diaspora the 'Facebook Killer'?" It had some good news to report.

"While the fact that new invitations are being sent out does not prove Diaspora will succeed or survive, at least they show the still-tiny Facebook competitor is more than mere 'vaporware.'"

With Yosem's departure any realistic hope of getting a beta out in November vanished. Dan was not troubled. The work had been hard, and good. They would get there. Organizationally, the project was in splinters, but there was fresh blood: T.H., the woman who had left Facebook and had introduced herself to Ilya at the Silicon Valley Human Rights Conference. On Wednesday, the ninth of November, she was a few days into her efforts to make progress. Someone new, from a real company, was giving them orders. A woman. The guys were glad to hear a fresh voice. Among her first tasks, T.H. imposed a regime that required them to clear up outstanding issues that had been submitted by users.

In the afternoon, she wanted each of the founders to say precisely what they were doing. One by one, each of the guys was pushed by her to explain where they were headed.

"Max," she said. "What do you really want to do?"

T.H. might have thought it such an unexpected question that they would be shaken out of their ruts and forced to think how their lives and goals meshed with the work they were doing. In fact, all of them had answered this question multiple times over the last few months, in one forum or another, and they could have answered for one another.

"I would love to be an entrepreneur," Max said, "and I want to be a software engineer."

She turned to Dan.

"What about you Dan, what is your aspiration?" she asked.

"I want to run my own business," he said. "I don't want to have a boss."

Max and Dan knew what was coming next. Ilya would get his PhD in math and become an expert in machine learning.

"I'm a programmer," Ilya said. "I've accepted, that right now, I'm a programmer."

The rhythm of this ritualized dialogue was lost for a nanosecond, the strangeness of Ilya's answer looping into Max's attention for an extra beat. In an instant, the conversation had moved on.

Dan was heading back to New York that day for the NYU panel. Before leaving, he had a word with Ilya.

"Take a few days off," Dan told him. "It's only a project. There's nothing to be ashamed about with depression. I see a shrink every week. I can give you his name."

Ilya thought that was a good idea. He would go next week.

"I really appreciate it," he said.

After Dan left for the airport, the only people left at Pivotal that Wednesday were Ilya, Max, and T.H. Around nine, everyone agreed it was time to go home. They put on their coats and were starting for the door.

"There's something I've got to take care of," Ilya said. "I've got to make a phone call. Go on ahead."

Max had no intention of hanging around to wait. It had been a long slog.

"We can wait for you," said T.H.

"No, that's fine, go on, this may take a couple of minutes," Ilya said.

More weirdness, Max thought. The guy had his coat on; why did he turn back? He and T.H. got the elevator down to Market Street, and headed on their ways.

Later that night, Stephanie Lewkiewicz, who had met Ilya two years earlier when he awoke her in the math lounge at NYU, came out of a UCLA library after a siege of study and work. It was somewhere in the midnight borderland of Wednesday, the ninth, and Thursday, the tenth. She had been working on her PhD. Her phone, which had been switched off, showed a missed call. It was Ilya. They were still easy enough with each other that he would understand if she did not call him right then and there. She'd get back to him in a day or two.

At NYU, Rafi, who had been dismayed by the turmoil and glad that he was distant from it, did keep his connections with the group at approximately the social level he desired. He had created a robot Twitter account under the name of the Hipster Guido, and it was set up so that every time Dan posted a tweet, the Hipster Guido retweeted it, preceded by the words "The King has spoken!" Dan took it good-naturedly.

Arriving in New York late Wednesday night, Dan had declared himself home by using Foursquare, a social-web application that tracked locations. It automatically noted that he and 114 others were at Kennedy Airport in New York. He tweeted a picture of the family cat, then added a grace note about his return to the city. Rafi's Hipster Guido robot passed along both the cat picture and Dan's note with the customary introduction:

> @hipsterguido:The king has spoken! @danielgrippi—lovingly made in new york.

With Dan on the other side of the country, Max decided he and Ilya could take their feet off the accelerator. Instead of going to Pivotal, they met at Max's apartment in the Mission on Thursday, the tenth. They did not get a lot done, and broke early. Ilya did not tell Max where he was going, but he had to make a stop at a party supply store to pick up an order.

On Friday morning, the eleventh, Max saw that Ilya did not seem to have slept. His color was off; it must have been days since his head had been beneath a shower. And instead of his regular machine gun of speech and ideas, he was speaking and moving in slow motion. He was wearing a T-shirt that said "Dawgma," the name of his high school robotics club.

It was the second day that Max and Ilya spent at Max's rather than Pivotal. A break from their routine, yes, but also a concession to reality. Even if Dan had stayed, they would not have finished, despite all the progress. They had raised enough money, about sixty thousand dollars from donations and parents, to keep going until January. The beta would come out then. There was no getting around the holes left by the departure of Yosem.

That was bothering Ilya; so was another problem, one that had long gnawed at him. He wanted people who used Diaspora to be able to delete material.

Suppose a girl posted a photograph of her with a guy, and they broke up, and she wanted to delete it.

Seemed simple enough. Hit the delete button. The picture disappears—at least from the server that it was deleted from. In a distributed system, each server was an empire of its own, not subject to the rules of a centralized hierarchy. That is, a single piece of data would live on multiple servers.

"It's really hard," Max said. "You have to go into all the servers and ask them all to do this operation."

Facebook said that for the most part, it did not continue holding things in its servers that people had deleted from their pages. But because Facebook kept data on multiple servers, all of which it controlled, any item might have to be deleted in its servers around the world, making the process less instantaneous than met the eye. It was like a letter or a picture stuffed in a drawer—until it was cleaned out, someday, someone might open it again.

Not surprisingly, Diaspora faced similar problems—the data was distributed far and wide—without the command and control that Facebook held. So they set a goal to have a deleted piece of data actually dissolved from the network within twenty-four hours. This delay troubled Ilya. It had been precisely what bothered him most when his classmates back at NYU had created a prank Facebook account in his name—that the fake stuff would linger on a server.

They had gone over this issue before. Ilya had taken on the technically complex job of writing the code to address it. It seemed that he could not summon the energy to focus.

Yes, it was Friday, but he was in a deeper slump than would come from simply hitting the end of the workweek.

"Let's take a look at the worst parts of the code," Max said, "and make it, like, the best."

Programming, he thought, was like writing. He tried to convey that to Ilya.

"In writing, you're making sure that there is some canonical piece of knowledge in one place; really good writing doesn't need to repeat itself. Clarity and brevity are the values—the most expressive thing in the least amount of space possible."

They moved deliberately through the coding. Max thought he might be teaching him something, but he realized Ilya was out of gas. He could not engage. Max was getting weary. At midmorning, they were joined by T. H. Nguyen.

They broke for lunch and went to a taqueria around the corner from Max's place. Ilya picked at a shrimp quesadilla. When they got back to the apartment, he was still dragging.

"Dude, you're really tired, you didn't sleep," Max said. "We'll break at three-thirty."

After all, Dan was not around to crack the whip.

"I don't know," Ilya said. "We have so much to do."

They beavered away for a while longer. At one point when Max left the room, Ilya confided in T.H. that he was very worried about their pushing the beta back another two months. People were counting on them. Max was back in the room.

"I have a crazy question for you," Ilya said. "Do you think we will ever really release the beta?"

Fatigue, Max thought.

"Dude, of course we're going to do it," Max said. "Why would you even worry about that?"

Ilya shrugged. They turned back to the work, and he seemed engaged. Then he leaned back.

They had scheduled a Meetup for the next day, a Saturday, at Noise-bridge, the hacker space in the Mission, to get feedback from users. In theory, Ilya had called the meeting, but it had been T.H.'s idea and she would run it. She was fresh. He was swamped and spent, and it was still only Friday.

He'd had enough.

"You know what, man, I am going to take you up on that offer to break early," Ilya said.

"Hey, dude, see you tomorrow," Max said, relieved.

"Peace," Ilya said. He took his bag and left.

# CHAPTER TWENTY-NINE

**M**ore than halfway through the panel discussion, the moderator asked a question that Dan could have spent the rest of the session answering.

"Do you guys have any stories of intrigue or betrayal that you want to share?" she asked.

The auditorium at NYU was full, though it was just nine A.M., a most unsociable hour for college students, on a Saturday morning, no less.

Onstage, sipping an orange juice, Dan did not show a flicker of interest in that particular query.

Instead, it was fielded by David Goldberg, who, with his brother, Ari, had started an online company called StyleCaster with the goal of bringing "style to the people." It was one thing to break up a partnership started with friends, Goldberg said, and entirely different with a family member. There are former friends, but not former family. "Shit is going to hit the fan. There will be moments at every stage and evolution of the business where it's do or die. Or at least in real time, it feels that way," Goldberg said. "Retrospectively, it probably wasn't that big a deal."

The program for the gathering had a biographical listing for Dan that went a little beyond his nerd credentials: "Since graduating NYU in June 2010, Daniel has been fighting the good fight, developing Diaspora full-time with his partners in downtown San Francisco. He's responsible for

Diaspora's good looks and product cohesiveness. His pants are tight and his V-necks are deep."

Dan had told the genesis story, and how they had changed, over and over, the workings of Diaspora until they screamed with doubt, no longer knowing whether it was any good. It was then, he explained, that their design tutor and guru, Janice Fraser, made them draw a triangle. At the base, the widest part, she had them write what they had started out with—privacy and freedom. At the tip was where people achieved those two values.

"In the middle is where it changes," Dan said. "She told us, 'Notice that your product has changed, like, about three thousand times in two months. What it's going to do, how we say it is going to work. The thing is, as long as you have that base, every one of those iterations of the product, you can just step back and say: it still goes off those core values.' And I think that's the most important thing you can do."

He and the others were applauded. There was no sense going over the brutal six weeks that had just passed. Who even knew what it all meant? The Grippis were waiting to bring him back to Long Island. Dan could not find his sunglasses. Rafi's robot passed along his tweet.

@hipsterguido The king has spoken! to the gentleman who jacked my sunglasses: you look like shit in wayfarers. sorry bro.

Dan might have been up early on Saturday morning for the event in New York City, but on the other coast, Max and T. H. Nguyen knew that they had to begin at a more sociable hour for their workday. They arrived around noon on November 12 at Noisebridge for a "Meetup on User Experience," also known as UX.

The agenda was to watch ordinary people use the Diaspora site and figure out what gave them problems and what they were excited about. Ilya had held one a couple of weeks earlier, and though it was marginally organized, it had given them useful feedback. This session would have a stronger, more purposeful flavor, as T.H. was now helping. She created a sign-up schedule on the Meetup website for one-hour slots, and invited members of the Diaspora group to bring along friends who were:

1. Average social media users (no early adopters, savvy tech people)

2. Female

3. Ages 13 to 25

She didn't need to write the words "No geeky boys."

She would videotape sessions of "normal people" test-driving Diaspora. Her experience at Facebook had given her a good idea of how corporate focus group studies work. Still, in announcing the session, T.H. made it clear that this was an initiative of Ilya's. She wrote: "Ilya envisions getting into a groove of doing user tests (every other Saturday), digesting learnings (during the week), then hacking to resolve surfaced issues (every other Sunday). Let's keep the ball rolling this upcoming Saturday."

They got an excellent response. There was only one problem. There was no sign of Ilya when they arrived at noon that Saturday. By one P.M., he still had not turned up. Max dialed his number, but the call went right to voice mail. He tried again a few minutes later, with the same result. People were coming for the first session, and although Max was annoyed, he could not give Ilya's absence much attention.

"Do you think he's okay?" T.H. asked.

Ilya had looked wiped out the day before, and Max thought, God knows, there are Saturdays when I don't want to talk to anyone, think about Diaspora, or do anything besides drop off the face of the earth.

"He's okay," Max said. "He's done this before."

Through the afternoon, Max sporadically tried Ilya's phone, but he didn't answer, and he and T.H. were busy with a full house through all three testing sessions. At the end of the day, T.H. posted an announcement.

"Thanks to all the people who let us pick their brains! We ended up having more users to test than video capacity to record . . . so we're canceling dinner to give people a breather."

Katie Johnson had just gotten a new job that came with a place to live in Pacific Heights, a tony part of the city, but most of her friends were still in the hipster parish of the Mission. She turned twenty-five at the beginning of November, and Tony Lai was hitting thirty. With so many friends in common, he invited her to cocelebrate their birthdays at the Hive on that Saturday evening, the twelfth of November.

They met at a Costco to load up on supplies, and then she headed back out to her house to get changed. Tony returned home. Festivity fatigue was settling on the Hive, which had been the scene of eight parties in a two-week span in October, including three back-to-back Halloween events. Ilya's birthday had been mashed in there. After tonight's gathering, the calendar would open up for a while and provide a revelry reprieve. Gardner and David were hanging out in their rooms as Tony unpacked. Ilya's door, which opened into the main living and cooking area, was closed.

A few blocks away, Max, who had no real connections with anyone involved in that night's Hive party, was making his own plans for the evening. A text arrived from T.H.

"Ilya's mom called me. He was supposed to call her four hours ago, and this is one of the last numbers he called."

Since Ilya's phone was on his parents' account, his mother had been able to look at the online billing record to see numbers of people he had spoken with recently.

Max answered: "He's ok."

T.H. replied: "R u sure he's ok?"

"Yeah," Max wrote. "He's stressed out. Sometimes he gets like this. No big deal."

"R u really sure? His mom just called me."

"I'll track him down," Max said, but he could not raise him, either.

To his surprise, Max did not have the number for David Kettler, the Hive roommate whom he knew best. He sent him an e-mail.

Around that moment, as Tony was unpacking the supplies, his phone rang, displaying an unfamiliar number.

"Hi, this is Ilya's mother," the caller said. "Have you seen Ilya? I haven't been able to reach him."

"Sorry," Tony said, slightly startled. He had never spoken with Ilya's mother, Inna Zhitomirskiy, or anyone in the family. "I haven't seen him for two or three days."

Tony said he would pass along a message for Ilya to call her when he saw him, and signed off. David was sitting on the couch, and Gardner was in his room. Neither had seen Ilya for a couple of days. Nor had Bobby Fishkin, who was arriving for the party. Behind him were more friends of the house.

"The party is going to start," Fishkin said. "Where is he?"

"We haven't seen him," Tony said.

"Have you checked his room?" Fishkin asked, simultaneously banging on the bedroom door and hollering Ilya's name.

"It's locked," Tony said.

Gardner felt his stomach knotting. They never locked the doors. He looked at the knob, which could be jimmied with a thumbnail: a hack. He pushed open the door, stepped into the room, followed by Bobby, Tony, then David. In a glance, Gardner saw all that he needed to, and turned around.

Fishkin looked at the bed.

"Ilya!" he shouted. "Wake up."

From the bay window of the bedroom, San Francisco sparkled and winked against the plane of a November night. Below, in the courtyard, stood the tent-cabin that had kept them warm on the chilly San Francisco nights. Notions and schemes had been shot from that third-floor window as if from one of those air bazookas that fire confetti at sports events.

The walls of Ilya's bedroom were lined with lists compiled after late-night bull sessions, composed on paper torn from the rolls of butcher paper. Next to the queen-size bed was a spindle, the kind of spike that restaurants use for paper receipts. This one held Post-its:

Go Skydiving
Start a company
Brush teeth
Sneak into a company and re-arrange everything in the file cabinets
Go to Burning Man
Live in a castle
Which Silicon Valley company is going to do the most to colonize Mars?
End bribery in Congress
Party with the Amish

Gardner shook Ilya's foot.

The laptop had been open on the bed, and one final Post-it note was stuck to it.

"Thank you to everyone who was kind to me. Please know this was my decision alone. Please post this."

Gardner eased the plastic bag from around Ilya's head. On the phone with 911, the operator wanted to give Tony instructions on resuscitation. He told her there was no point.

At his parents' home on Long Island, Dan had watched texts of people trying to find Ilya flying across his screen. Then his phone rang, lit up with Max's number. In the instant before answering, he had a single thought: Fucking Ilya had offed himself.

Ilya had been dead all that sparkling Saturday in San Francisco, as his mother and father and little sister went about their lives in Lower Merion Township, Pennsylvania, and his partner Max ran the Diaspora users' tests in Noisebridge, and his friends bought provisions for yet one more party at the Hive.

He was dead as Dan spoke that morning on the other side of the country to the audience at New York University about the joy of building something.

They were still doing Diaspora, Dan had said that morning, because people wanted it. As many changes as the project had gone through, to whatever effect, they always rested on the core values of privacy and freedom.

When they started out, he said, "We never saw it as a start-up, necessarily. At the time, we were four geeks, wouldn't it be cool to make this distributed platform?" And then, he said, with the flood of money, they realized: "People really want this. We have to make this thing for the Internet. People are giving to the project because they believe in what we are doing. It's kind of wild. That's why I say, it's not the real world."

The moderator had asked about that very subject—on when the real world began for young people, the sails of their lives billowing with ideals and ambitions and things that had not been done or dared before.

"I went from school, doing something that I love, to, like—doing something else that I love!" Dan had said. "Yeah, you trip up sometimes. It's a real rush."

"You're telling us the real world hasn't begun yet?" she'd asked.

The audience laughed. Dan smiled.

"Ahhh," he had said. "Yes."

He and his parents, Carolyn and Casey, stayed up all Saturday night. Sometime after midnight, Dan posted a tweet.

FUCK NOTHING AND EVERYTHING.

He and Casey took the first flight out of JFK to San Francisco in the morning.

Before the investigators from the medical examiner's office left the apartment with the body on Saturday, they told the roommates that Ilya had purchased the tank of gas that he had used to suffocate himself on Wednesday. They apparently had been able to get his credit card records. Max recalled his strange start-and-stop behavior leaving the office that night, when Ilya, T.H., and he were all ready to go and Ilya pulled back, telling them not to wait for him.

He had picked up the tank sometime on Thursday, along with a package of balloons, which apparently were thrown in as part of the rental. The balloons were untouched on his desk. On Friday at Max's house, he agreed with his suggestion that they knock off early for the day.

No one had seen him since he left Max's Friday afternoon.

On Saturday night, Elizabeth Stark, who had been eating in a restaurant a few blocks away, arrived at the Hive soon after the body was discovered. Elizabeth and Ilya had had a platonic friendship of rare power; she had been a guest at the Hive on a number of evenings when the talk had stretched too deep into the night to return to Palo Alto, where she was teaching. They had shared his bed companionably. In crowded rooms, each would know where the other one was.

No surprise, then, that even though Ilya had been the cryptography czar for Diaspora, Elizabeth figured out the password to his laptop in a minute or so.

Ilya, who had often joked about his tinfoil hat affectations, frequently used anonymizers when browsing the web that disguised his location and identity. That made it hard for anyone or any machine on the other end to know who or where he was. But his laptop had no such obfuscation. His browsing history was spelled out in unmistakable detail. For months, he had been looking at sites that discussed suicide techniques. A few minutes after nine on Friday night, he had downloaded a PDF document that gave the instructions; he had already downloaded the same file several times that evening.

Just after eleven P.M., he telephoned a woman he had spent some time with, but she was in her car and didn't answer.

The timing of the call struck David Kettler, the roommate who had moved into the Hive that summer after being critical of Diaspora during a Stanford field trip to their office.

That Friday had a peculiar spell over mathematicians, and many members of that tribe, including David, had been paying attention to its approach. Friday was November 11, 2011, or 11/11/11. For those beguiled by numbers, it had the distinction of lining up the month, day, and year in 11s, the lowest two-digit prime number. It would not happen again for a century, and would not fall on a Friday for another four hundred years.

The last call had been just after eleven P.M. Friday night. To David, it was beyond question: Ilya had waited until the minute, hour, day, month, and year were all lined up on 11, then drawn his last breaths.

Someone taped a sign to the door of the Hive saying that the party had been canceled.

On the other side of the country, in Cambridge, S. J. Klein had gotten the word from Elizabeth Stark. Rather than sleep, he replayed the night a few weeks earlier when he and Ilya had stayed up until dawn talking, right after they'd met at the Silicon Valley Human Rights Conference. And Klein also could not get a video that he had just seen out of his head.

It was made by people in a rowboat on the River Shannon in the southwest corner of Ireland. They had gone out to document starlings in a flock: a vision of tens of thousands of small birds, swooping in unison, the dark clouds of their mass splitting, then suddenly rejoining. A funnel of life spread across miles of sky, spilling out of formation and back in an instant. Scientists could guess what was happening, but no one really could explain why the birds did all this. It was called a murmuration.

At 3:16 in the morning, Klein posted the video of the murmuration onto his blog, beneath the headline that could have come from a tombstone: ILYA ZHITOMIRSKIY 1989–2011.

"Together," S.J. wrote, "we can move in ways that none could move alone."

# CHAPTER THIRTY

**M**ax poured the day's first shot of coffee into an Elvis Presley mug from the "Hound Dog" era, then went out to the garage where they had furnished a space with folding tables and a couch they'd bought off Craigslist. They held their group meetings there, gathered around an exercise bicycle, and did their work at tables that looked out to Gloucester Street, a quiet cul-de-sac in Sunnyvale, one of the suburban towns that make up Silicon Valley. The town was rightly named. Every day had been bright, warm, comfortable. For three months, the reconstituted Diaspora team of him, Dan, the designer Rosanna Yau, and a programmer named Dennis Collinson had been renting a house there.

It had been just about ten months since Ilya died, and Max had called a Diaspora community meeting for that morning, Monday, August 27, to be held in an Internet chat room.

"We will be making an important and exciting announcement about the future of the project," he had written.

When Max logged in, one person was already in the Diaspora meeting room with a question.

"Is there going to be music while we're waiting?"

Max quickly cued up a song that had been popular almost twenty years before he was born, but was right for the day: "Monday, Monday" by the Mamas and the Papas. Once it finished, he pulled up another pop hit from the 1960s, by the Monkees: "I'm a Believer."

Few people at the meeting would have sung along. To all but a small group of its most devoted followers, Diaspora seemed to have withered in the months since Yosem had left the group and Ilya died. The four of them were forty miles south of San Francisco, nowhere near where they had started.

For five nights after Ilya died, Max opened his place in the Mission for a "shiva-fest," a variation on the custom of Jewish mourners receiving visitors over many nights. They cooked and ate and drank. They held on to each other. Casey Grippi shopped. One evening, about twenty people sat and sprawled in the living room, staring into their phones or tablets or laptops, most of them saying nothing. But small smiles broke out. With no planning, no announcement, they had started texting and tweeting to one another. Casey showed up with twenty-two cans of tomato sauce in honor of Dan's twenty-second birthday. A hashtag was established for their tweets: #saucecon, as in, sauce conference. The physical presence of people in the room could not compete with the screens where their bon mots were popping up. Max and Dan hacked up a program that let them modify and send GIFs. So a cat could appear to rumba; a big dog could jump in the air at the sight of a tiny mouse. Their little hack made it possible to add silly captions that they could send zipping into cyberspace and onto the screens of the people sitting at the other end of the couches.

Finally, on the Friday after Ilya's death, it was time to move the ritual forward. Dressed in a suit, his body lay in a casket at the front of a San Francisco funeral home. Elizabeth Stark set up a live stream so that his parents, who were in Pennsylvania awaiting the return of the remains, could watch the services. Max opened the program, speaking of his friendship with Ilya and inviting anyone to speak, just as he had opened his home in the previous week. A dozen or so people spoke of Ilya and his enthusiasms, his support for the improbable, his deep empathy, his visions for a better world. Aza Raskin compared him with a Swedish reformer who modernized her country's impenetrably difficult traditional alphabet simply by insisting that all books for schools be printed with a modern, simpler one. In a generation, she changed the practice of centuries. Ilya's legacy would be like that, Aza predicted.

The only direct mention of Ilya's mental health problems was by

Mitch Altman, who had cofounded Noisebridge. At fifty-four, he was an elder statesman of the tribe. He also had invented TV-B-Gone, a device that Ilya had before he ever came to San Francisco. Altman said he knew personally how bleak depression could be and urged people to look out for one another. He felt that he had failed Ilya.

Dan sat toward the back, neither speaking nor looking into the coffin; he doused his anger with silence. Yosem Companys, receiving emergency treatment for an asthma attack, had been unable to come.

Two days later, the Zhitomirskiy family held a service at the First Unitarian Church in Philadelphia, a historic congregation founded by Joseph Priestley, who is credited with the discovery of oxygen and the invention of soda water.

The Zhitomirskiys were joined by, among others, the other three Diaspora founders and their parents, their mentor, Evan Korth, and NYU friends. Rafi gave the most concise speech: Even when Ilya was angry, Rafi said, he did not want to push people away. So Ilya smiled. Rafi intended to keep that in mind. Max repeated his talk from San Francisco; Dan encored his silence. Afterward, Dan drove home to Long Island with his parents, and Max headed to Washington, D.C., where his mother and father had recently moved.

A week later, Max and Dan returned to Pivotal. The Diaspora code forge had not shut down. While they were away, their friends at Pivotal had assigned two programmers to pick up where the guys had left off, working through a stack of bug reports that had piled up. The whole office had been shaken. People were glad to have a way to pitch in.

One of the Pivotal people could not bring himself to return to the office. Dennis Collinson, an intense twenty-three-year-old Matt Damon look-alike, had flown from San Francisco to Philadelphia for the services, and then stayed with friends in Brooklyn. His bosses were fine with him working at the Pivotal office in the Union Square neighborhood of Manhattan; it was as easy to do the pair programming across the country as it was across a table on Market Street.

Dennis, though, found it harder to focus on the work. He was drinking heavily. He liked Ilya, considered him a friend, but they were not bosom buddies; he went to the parties at the Hive, but did not find them as captivating as others did. Even so, his death had knocked Dennis out of the pleasant haze of a life that, he said, consisted of "orgies and brunch."

In life, Ilya had made it possible for him to work in the corporate environment. Once or twice a day, Dennis would swing by Ilya's desk to talk for a few minutes; he and Ilya would take a night to go drinking or play board games at Mike Sofaer's house. "This dude is telling me about freedom," Dennis said. "Socially, I was very, very free, in a way that he really couldn't understand. But globally, politically, I had given up." As glancing as their friendship seemed to have been, he did not think he could face Pivotal again without having a few minutes of Ilya during the day.

Around the time he turned sixteen, Dennis left his home and Catholic high school in New Jersey, and entered a world of underground gay parties mounted in old warehouses and factories in Brooklyn. He would be gone for days at a time, not bringing his computer with him, though he had made his first website when he was eight and stopped using the Windows operating system when he was thirteen in favor of the free software Linux. He knew that he could disappear into the computer for hours. Joining the Radical Faeries—"a loosely knit spiritual network of crazy faggots," he explained—he found a society that mentored and supported him as he learned programming and lived exuberantly. In time, he settled in San Francisco.

At the end of his workdays, he went home to a cottage at the bottom of Bernal Heights, a San Francisco neighborhood that was becoming hot. The bike ride to work was flat, a prime virtue in a city of stupefyingly steep hills. His rent was twenty-three hundred dollars a month, but the view was fabulous and, set back a hundred feet from the road, his home was a retreat.

Join Diaspora and give it up?

In his grief, he had been doing a crummy job for Pivotal, as he knew well. He had about four months of savings. There was still a shot to make Diaspora happen. He had long thought that there had to be a better way to do social networking than Facebook. Diaspora was a way to actually make it happen. But before Ilya's death, he had been unable to find a way into the project; the Diaspora guys had formed a tight four-sided box.

"I could maybe write better code from scratch than where Diaspora is now," he mused over drinks that winter, "but I suck at making community happen, and they have a better community. And they have the attention of the world. And they're my friends. I love these people. I want to see this keep happening. I am never going to make this happen on my own. I have the power to help turn this light back on."

In the haze of those winter months, Dan and Max could not resist his help. They agreed to take him on, and Dennis walked away from a comfortable life.

"I don't really have any other choice," he said. "If I don't, it's something I would regret for the rest of my life."

He was not the only reinforcement they would get in the months after Ilya died.

One night, as the group wandered through the bars in the Mission, Rosanna Yau turned to Max. Rosanna was the artist and designer who had come to them during their first summer in San Francisco, offering her services, and the four guys had struggled to figure out what questions to ask during her job interview. Then they decided that they could not spend the money, but she had become enchanted with them. She had come up with their milkweed logo. She had cooked bugs for Ilya.

Now, in the dark winter weeks after his death, she had committed to the project, and was with Max, Dan, and Dennis several days a week.

Basic questions kept turning over in her mind, almost too blunt, too cosmic, to ask during the normal workday. But they were relaxed now, a few drinks into a convivial bar scene. No better time.

"So Max," she said. "What is the future of Diaspora?"

He sighed.

"I can't answer that question at two in the morning," he said.

# CHAPTER THIRTY-ONE

Dan was pretty sure he knew the future of Diaspora, but he could not bring himself to utter the words. There was someone in his life, however, who had no such inhibitions.

For months, Dan had been seeing a psychotherapist on a weekly basis for anxiety. To the world, Dan was a handsome, smiling face, a man of few words beyond a sardonic crack here and there, his delphic tweets about arcane changes in programming languages, the music mixes he created, and a stream of pictures that he captured from the Internet. Like Max, Rafi, and Ilya, he was careful to avoid speaking ill of Diaspora in public. He was truly joyous about the banquet of possibilities that life presented, as he had told the NYU conference twelve hours before Ilya's body was discovered.

In his therapy sessions, though, another side poured out. The meetings with the venture capitalists had terrified him: their group careened from one pitch to the next, changing its story, never coming up with a coherent way that an investment would be paid back, or, indeed, mapping a path to survival. Or explaining why, in the end, anyone should even bother with Diaspora. It sucked. He didn't use Diaspora. He saw no reason to.

As the weeks went on, Dan heard the therapist form questions from the shapeless unhappiness that he detected in his patient.

"Why are you doing something that you don't want to do?" the therapist asked. "Every week, it's like you're coming in here and you've been kicked and you've been stomped."

This was true, Dan realized.

"Why don't you leave the group?" the therapist asked.

That thought had crossed his mind before, but the idea of walking away made him uncomfortable, not simply because it would have meant failure but because he felt that the value of the core idea—its social and economic worth—had not been mined.

"I know there's something here," Dan said.

"It's over," the therapist said. "Move on. Accept failure."

After Ilya's death, he was flooded with the doubts that his partner's enthusiasms had, somehow, kept dammed up. The arrival of Rosanna and Dennis had brought a burst of new energy in the first weeks of 2012, with Dennis speeding up the response of the program, and Rosanna bringing a fresh eye to the design. And the long sieges of work had broken. The Daytime Emmys, the award show run by his mother, Carolyn, was being held in Las Vegas, and the other three Grippis, Casey, Laura, and Dan, joined her there.

Still, he was treading water. So was the project. They had enough money to last another four or five months, at the most. It wasn't just cash that they were burning up: Dan felt they were imaginatively bankrupt, that the gusts of enthusiasm from the new Diasporans would not get them anywhere. They didn't know where they were headed. A sense of duty kept him going to work. He worked with Rosanna on the design, but they made only sporadic progress. Maybe, she suggested, the schedule that brought her into the office only two days a week made it hard for them to get things going. Although she could not cut back on her other obligations, she did have time on the weekend.

Rosanna arrived at Dan's loft in the Mission a little bit after three on the afternoon of Sunday, February 12. The professional life of his photographer roommates was everywhere visible: big light umbrellas, backdrops, screens. Daylight washed the living room, which was beneath a glass roof. Dan pulled a table out and they sat down.

Their laptops stayed shut.

"I have no idea what we are doing," Dan said. "What is the point of all this?"

They had been moving features to different spots on the Diaspora web pages—adding hashtags here, a new drop-down box there, subtracting and adding and trying to figure out what would entice and engage the users.

"We're just pushing pixels around," Rosanna said.

For each element of design in Facebook or Google, the giant social networks could get advice from platoons of marketing researchers and cognitive scientists who could offer guidance on what they were doing. Whether an individual feature was good or bad, beautiful or weak, the gravitational force of hundreds of millions of users made the giants inescapable, like dark stars. No Diaspora feature, no matter how clever the gut instinct from which it arose, could match those forces. When Dan and Rosanna had started collaborating in January, he initially called their work a redesign. Then he changed the term to "rescheme." That Sunday, he was more dismissive.

"This is putting lipstick on a pig," he said. "A waste of time."

In this case, the lipstick was the user interface, the UI, the bag of visual and interactive tricks offered on the site.

Rosanna agreed. "You can move all this stuff. Anyone can make awesome UI. All this shit can change. We are trying to reinvent the wheel," she said. "If you put a feature out, tomorrow Google is going to make the same thing."

They had only to look at how Google came out with circles months after Diaspora had introduced the aspects settings, each of them a digital corral, a way to hold aunts and uncles in one place, girlfriends and boyfriends in another. Maybe Google engineers had been quietly dreaming up their version of it even before Diaspora went public, or maybe they had not. Either way, it was a perfect example of how quickly digital innovation could lose its novelty. Technology could be replicated, just like the MakerBot had been whelped from the idea of early 3-D printers.

Dan needed a cigarette.

"Let's go for a walk," he said.

The streets of the Mission on weekends are not quite the class 5 white-water rapids of Times Square, but they do have respectable rivers of walkers. The coffee shop down the block from Dan's loft was chronically crowded; even if they could find a place to sit, having a conversation in its din was like trying to talk above three armed riots. They walked. Dan lit up an American Spirit cigarette. Rosanna turned back to their first principles.

"Your vision is what holds you together," she said. "The thing that no one can touch you on is that you are decentralized. Google and Facebook are already deep into their own structure. How are they going to make money off that? You guys are the one. You have the voice. You are visible. You almost have the responsibility to move this movement forward."

Federation, distributed networking, decentralization—like data ownership—were things that few normal people gave much thought to, Dan believed, although he knew it was a geek cri de coeur. "Decentralization is a nerd boner," he said. The deeper question, though, was what it could give people. For Dan, the answer was plain: the ability to express themselves, to connect without corporate boundaries. AOL had been ubiquitous in Western culture through the 1990s, until people realized that Netscape and Firefox had made the web a place they could navigate unsupervised. Once it seemed safe to roam outside the garden walls of AOL, that company's subscribers evaporated into cyberspace.

"We keep saying that Facebook is going to be the next AOL," Dan said. "We never said what the solution is going to be. Oh. We're going to put a search bar at the top, search for hashtags. That sounds a lot like everything else."

Rosanna recalled that at her first meeting with the group, the laughably awkward late-night interview in the Pivotal offices, the guys had talked about creating a project for girls, not the beards. Something for everyone. That ambition, so striking in its earnestness eighteen months earlier, had not come close to being realized. Anyone, it was true, could join an existing pod, but setting one up was a job for a full-grown beard. Diaspora could be downloaded by anyone for free, and installed on a server, but it demanded fluency in arcane details that most people had no time or inclination to acquire. Making it easy to install was for some unspecified time in the future. That meant even the installation would be a barrier.

Dan had just finished reading a biography of Steve Jobs by Walter Isaacson. "What Jobs did was make complicated things simple," Dan said.

The sun was dropping as they walked back to his place. Upstairs, they continued to talk. Simplicity was a prime virtue. They had to make it—whatever it might be—easy to install.

But a deeper, more profound change was needed. The whole idea of Diaspora as a place was wrong. The very notion that users needed a networking "site" for their thoughts, comments, updates, pictures, was misguided, a distraction fostered primarily by Facebook's setup. Yet Diaspora had bought into that notion. It had created tightly scripted boxes, just like those of Facebook and Google and Twitter. Why, if they wanted to, couldn't the users spread a picture, a piece of art, a single word, across the full screen as their profile page? Why should the social networks inhibit people's expressive abilities?

Of course anyone could set up a personal web page to their exact design, and millions of people had, to virtual silence. "They're dead zones," Dan said, networks of one, static presences that lacked the voltage of interactivity.

Then there were blogs. WordPress had spun the wheel of progress forward a complete turn. "It's great for nerds. It's great for the average person," Dan said. "But you can't interact with them."

Inevitably there were elaborations of WordPress—Tumblr was a microblog tool created by David Karp, a twenty-one-year-old who had left high school in New York after his freshman year—but it, too, had limitations, being made with proprietary software.

"The personal website is dead space," Dan declared. "But make it connected. Match WordPress with the coolness of being connected to Facebook and Twitter."

He pulled out a notebook and began to draw, ink on paper. Everyone would have a personal web page, with all its freedom, and it would propagate through the existing networks. Diaspora was lonely: no one went there except those who thought about Diaspora.

Suppose, Dan mused, that you could tell your friends on Facebook and Twitter that you had just updated your own web page. "That's how we beat the network effect—piggyback the shit off Facebook and Twitter," he said.

Now Dan spoke at length, wildly unusual for him, simultaneously drawing his ideas in a notebook. The power of democratic publishing, made possible by WordPress, would be amplified by social networks. No one was doing that. He and Rosanna began to think aloud about how it could work. Some people would host their own pages, but there would be a demand for a hosting business. Companies would like to have the ease of a flexible web space that they could modify with the ease of anyone updating a blog—and still have the vibrancy of social networks. "If my mom is putting on the Daytime Emmys, she can sell tickets right on the page," Dan said.

There were clear points of revenue. It gave people total control over their data. If they wanted to sell it to advertisers, fine. If they didn't, that was fine, too.

Finally, Rosanna thought, a plan.

"We just have to convince Max and Dennis," Dan said.

"You can pitch them tomorrow morning," Rosanna said.

"You have to have my back," Dan said.

"I will," Rosanna promised.

It was a matter of do or die, as far as Dan was concerned.

Dan called for a morning meeting, cracking his timekeeper whip. He'd kept working Sunday night after Rosanna left, trying to find the weak joints in his plan. There was nothing that could not be fixed. Now he had to find out if the others would go along with his ideas.

"Guys," Dan began, "I'm going to be looking for a job in three months."

"Okay," Max said. "Sure."

"What we are doing right now fucking sucks," Dan said. "The product is shit."

He recited what he saw as the implicit, unspoken demand of the project as it existed: "Join Diaspora, we are cool guys, so it's totally fine." No one tried to argue the case. Of the four people in the room, Rosanna realized, she used Diaspora more than Dan, Max, and Dennis put together.

Not that Dan needed to be more blunt, or could have been, but he saw a way forward. He was confident in it. They could try it or not. If they didn't, he was done.

So he laid out his plan. Rethink what it meant to be in a social network: not people adding a pinch of update here, a cute photo there, a link here—whatever class of items was permitted by the corporation that ran it—but a sharing of personal websites, which had the most scope for personal expression, and which currently lacked the vibrant charge of social connections.

For a moment, he discussed the possibility of creating a pay service to host the websites. "Say Lady Gaga has her shit there. We say, 'Given your load, we can off-load a hundred thousand uniques a month.'"

Dennis did some arithmetic, comparing the pop star with their most ardent young user in Amsterdam.

"Kevin Kleinman can be charged nine dollars and Lady Gaga a hundred thousand dollars," Dennis said.

A delightful thought.

Dan returned to Kevin Kleinman.

"He doesn't want that much," Dan said. "He can't have a custom theme on Facebook. He can't own his data on Facebook. If he wants to do that, he has to have his own personal website. And he doesn't know how to do that. If he does, he has very limited knowledge. Once he does that, then he doesn't get the comments and interactions. That's what Facebook gives him. We allow this comment and interaction lane. That's it. Supersimple."

There simply was no reason to feed and care for the existing setup. "People go to JoinDiaspora.com and they say, 'Great, another social network. My friends are not here. Who cares?'" Dan said. "We are going to make it dead simple for you to make and manage a website. You own the data. And it can become social."

Dennis and Dan talked about how they would explain it to users, and investors. Dan said: "WordPress made your personal website a blog. That was cool. You could just track WordPress. We're going to do the same thing. This has utility—people already have websites, minus the social factor. You tie in the social factor, and it's a million times better. It's not like, why am I signing up for a new service. It's like, I have a website and it's socially enabled. Right now to make a website social, you need your Facebook fucking like button on it. And it all goes back to Facebook.

"Your personal website should not be your Facebook profile." With this new approach, users could do what Facebook and Google did with their data: auction it off.

"You have your own website, you tweet about it, you get more people to sign up. You get your analytics. You get your payoff. It's really pretty simple. You own your shit. You go from there," Dan said. "It's what people want. WordPress proves the business model."

Max, who had raised a few objections to the plan, balked at invoking WordPress. "Every time I ever mentioned WordPress to a VC, they were always like, 'Okay—they don't kill it,'" he said.

WordPress was not, in other words, a monster generator of revenue, nor did it have the prospect of becoming one, even if it was a dominant presence. "For being six out of ten website installs, they are not killing it," Max said.

This puzzled Dennis. It was one of the most important tools to have emerged from the twenty-first century. "How are they not killing it?" he asked.

"They make enough money to be a sixteen-person company," Max said.

This did not trouble Dennis.

"Maybe that's what we are going for," Dennis said.

Dan said, "They're not killing it because they are not social and everything is social."

"That's fine," Max said.

They could not, in fact, sit around much longer debating the perfect approach, the best website.

"We've got four months," Dennis said, calculating his savings.

"I'm applying for jobs in three months because I will be flat broke," Dan said. "What we have right now, it's not going anywhere."

They had just burned up a month changing background colors, making avatars with rounded corners, trivial pursuits.

"We can't keep pixel pushing," Dan said.

Rosanna had mostly been listening since it was evident that she was not needed to back up Dan, who was making the case and had Dennis on his side. But she made it clear that she, too, was through with pixel pushing.

"I don't want to," she said.

"Rosanna doesn't want to pixel-push. I don't. Like, I changed the mobile site. Nobody gives a shit," Dan said. "We need to do something that's completely different. Twitter can change their design. Facebook can change their design. They can't do this. Nobody can do this, except for us."

Dennis sketched out work plans for the next week, including dealing with a persistent porn poster on the JoinDiaspora website. Dan said he would take care of it—after they ate.

"I forgot to eat breakfast and dinner," Dan said.

"It's lunchtime," Max said. "Why don't we just eat?"

"I'm hungry," Dennis said. "Hella hungry."

# CHAPTER THIRTY-TWO

They looked through all the name tags, about 150 for every start-up except them. One that had a plan to sell empty seats on airlines. Another that was making an app to record game playing on mobile devices. An app that would connect truckers with shippers, bypassing brokers. A custom-sizing program for T-shirts.

Nothing for Max or Dan, Dennis or Rosanna. Not one person in their group had a tag.

"Oh, *you're* Diaspora," the administrator said. "We heard all about you last night!"

It was early May 2012, and the Diaspora team was in Mountain View, California, with the summer class of start-ups at Y Combinator, the incubator in Mountain View, California. Ever since February, when Dan had his epiphany about making Diaspora a social network of personal web pages, they had been going guns blazing.

One evening in March, Max brought Dan to dinner with a friend named Jessica Mah. She was a prodigy entrepreneur who had started her first business at thirteen, graduated from high school at fifteen, and at age nineteen created inDinero.com, a thriving bookkeeping, tax-filing, and payroll business for start-ups and small businesses. The Diaspora saga was well-known in the Bay Area; if more than a few of their contemporaries took thinly disguised pleasure in their being brought to earth, Mah was not among them. She knew Max and liked him. And two years down a path of failures, misfortune, dumb mistakes, self-inflicted injuries, and naïveté, they were sitting in her living room, talking about how they would make it work.

"You should apply to Y Combinator," Mah told them.

Y Combinator start-ups came to Mountain View for three months, got some money and mentoring, and at the end of the semester presented their ideas at Demo Day, a beauty pageant attended by hundreds of investors.

Mah herself had been through the Y Combinator a few years earlier. And the Diaspora crew knew it well. It was the same program they had begun to apply for two years earlier, back in 2010 when they were still at NYU, but had given up on because the process was too much of a hassle. Max had actually read about it when he was eighteen, seven years earlier, and decided then that it was just what he wanted. At that point, he did not know how to write code and did not have a start-up idea, but he was excited by the possibility of creating a business. He quickly spun up an idea to create a web app for taking notes. He was not accepted.

It was a mark of prestige to spend a few months being mentored and measured by entrepreneurs and VCs. In fact, bringing in companies that were on the verge of collapse was a recently developed specialty of Y Combinator. The most famous of those was Airbnb, started by two young guys in San Francisco who could not meet their rent, and in desperation decided to advertise to attendees at a design conference that they could accommodate a few people on air mattresses. They had discovered a business linking convention visitors with residents who had extra space. They imagined a market for leisure travelers looking for accommodations that were bargains or less antiseptic than hotels. To raise money, they custom-made cereal boxes of the 2008 presidential candidates. For forty dollars, people could buy a box of Obama O's or Cap'n McCain's. This brought in more than thirty thousand dollars. It was enough to keep them going until they got into Y Combinator, where they patched together real money for a successful global business.

Ten minutes into the conversation with Mah, both Dan and Max realized they should apply. It would mark them, once again, as full of promise, not the fools who failed to kill Facebook. If they got into Y Combinator, over the course of the summer they were all but guaranteed about $175,000, in exchange for shares in the company. With Dennis and Rosanna, they quickly knocked out the application.

The interview was a disaster. In the car on the way down, Dan tweaked the coding for the new photo uploader on their demo pages. It didn't work when they set it up in Mountain View. The new Diaspora couldn't run.

Could they explain the point of the product in a sentence?

Max said he didn't think that was possible.

Dennis chimed in. "We want to help people make things and share them," he said.

That view resonated with one of the Y Combinator partners, Garry Tan, who was not much older than the Diaspora group and was interested in the evolving social media world. He had been a cofounder of Posterous, a clever blogging platform that had been bought and folded by Twitter.

"This is cool," Tan said. "You guys are cool."

None of the other interviewers showed enthusiasm. Certain they had crashed again, the group left Y Combinator and treated themselves to (fully clothed) massages in a strip mall, a twenty-dollar indulgence that was an occasional treat and team-bonding exercise. Right after that, they got an e-mail from Tan, confirming what they already knew: they would not be accepted.

They continued on with their work anyway, realizing that their cash was draining away.

One Sunday afternoon, Dan ran into Dennis in the locker room of a climbing gym. Dennis wanted to know why Y Combinator had turned them down. In his message, Garry Tan had not spoken about their dreadful presentation. The problem, he said, was that their corporate structure was so tangled, it was hard to tell where an investor could fit in, and what shares were available.

"That reason is bullshit," Dennis said. "You can reallocate the shares. That's not a reason not to fund it."

"Totally," Dan said.

"Why don't you write to Garry Tan and tell him that?" Dennis asked.

Dan agreed. Why give up? But Max said it would look like they were making excuses for their poor performance in the interview. Dan was adamant.

"We were hoping you would ask us about that," he wrote to Tan. "We had that answer. We can totally fix it."

There was no reply for a few days. Then Tan called.

If you're going to do this, he said, you have to do two things. You have to listen to what we say, and you have to fix the structure.

They had a deal. But they still had to write their own name tags when they arrived for orientation the next day. They had been added to the class so late, no one had had time to print them.

Nearly two years to the day that Max and Dan had graduated from college, the morning they had appeared on the home page of the *New York Times* and their Kickstarter campaign had erupted, *Bloomberg Businessweek* published a long profile of the group.

"In about a year, they had successfully built a social network that functioned more or less like Facebook but let people own their data," Karen Weise wrote. "Without any real marketing, more than 600,000 people used the site, but it was a narrow slice of the population, attracting lots of distrustful techies and Europeans."

That a journalist was there to praise them, and not bury them, was astounding. By then, any article about Diaspora could have been its obituary. It should have been a chapter, or a page, in the encyclopedia of Silicon Valley failure. Half the original team was gone, with Rafi back in school and Ilya dead. The project shot like a comet through the venture capital wings of Silicon Valley, but flamed out. They had lost a CEO who believed in their mission and had seen to forests of details. It was true that they had long ago kept the promise they had made on Kickstarter: to build a code base and get it out to the open-source community for hacking. Still, as of February 2012, just three months earlier, Dan acknowledged openly that he never used Diaspora outside of work. No one in his non-nerd life did. Max's feelings were complicated, but he himself turned to Facebook for the bulk of his social networking. By all rights, the last days of Diaspora should have come and gone.

Yet out of all the failure, they had come up with wisdom that no one else had. Personal websites "were an outlet for true self expression; they were what the internet was made for," Dan wrote in an unpublished essay. But, he said, they were not built for feedback, for dialogue, for comments.

They had the charge of human expression, but lacked the voltage of social connections. "Even now," he wrote, "the personal websites of the Diaspora Core Team are desolate places.

"Loneliness killed the personal website."

The idea of the new Diaspora was not just data ownership but creative ownership, too. There was no point in making a clone of Facebook that sat atop code gears that happened to be called "open source." Diaspora might just as well be Facebook if it was going to dictate to people how large their pictures could be, or that all emotional states, all thoughts, had to be captured in identically sized Lucida font, rendered in blue and white: the songs of the world sung in a single visual key.

A page on Diaspora would not be a regulated space that limited what and how the users spoke. Dennis and Dan were rewiring the Diaspora site so that a user with a lot to say could display comments in three columns, like a newspaper; someone with just a few lines would be able to render it in minimalist type. The *Businessweek* writer clearly liked what was happening: it was fresh thinking, a second idea, of creativity flexibility, atop the data ownership and privacy that first drove Diaspora.

"We are giving you both emotional and technical investment in what you are creating," Max told the reporter.

There was a great, hopeful development, *Businessweek* reported: "If anything can help Diaspora fulfill its ambitions to make a popular product out of its alluring ideals, it's the startup accelerator Y-Combinator."

It made perfect sense. Except that the first thing that Y Combinator did was virtually shut down Diaspora.

At the beginning of June, Max and Dan, Rosanna and Dennis, all moved into a rented house in Sunnyvale. Their first order of business was to outfit the garage as workspace with furniture from Craigslist.

The second order was kicking Diaspora to the curb.

That day, they had their first session with Paul Graham, start-up guru and creator of Y Combinator. Within a few minutes, Diaspora would come to an end, at least as a start-up.

At that meeting, Graham said they should put Diaspora aside, and concentrate on perfecting a set of tools—playthings, really—that they had developed in the days immediately after Ilya's death. At the saucecon gatherings in Max's apartment, visitors were sharing increasingly foolish texts and tweets; someone hacked together a way of creating GIFs with captions. A natural, organic network of jokes emerged. It was a kind of meme-generating process. The new version of Diaspora had incorporated these tools, but Graham said they ought to be stripped out of it. Then the team could spend the summer perfecting them, and putting them on a web platform where users could test them. Maybe the tools would amount to something on their own.

Graham strongly advised a clean break from Diaspora. They called the new app Makr.Io. The new web page was up on June 4. A press release went out the following day: "Users on Makr.Io can combine photos and text to create funny, unique posts, which they can invite friends to remix. A 'remix' simply allows users to change the text or photo of another post to change the context and give it their own meaning."

In an e-mail to me, Max explained: "Trying separate server to see how people like new ideas separate from d* (although code is 100% d*). Paul Graham's orders."

It was the start of another excruciating summer.

By the end of the month, there were quarrels over what they should be doing, and Dennis, the newcomer, was caught between Max and Dan. On the evening of one especially stressful day, he dug up a camouflage-patterned skirt that he had brought to Sunnyvale. It was a relic from the early days of the gay liberation movement, when someone had the notion that by feminizing war making, wars would come to an end. Dennis put it on and went for a walk in the evening air with Rosanna. They had not gotten very far when a car rolled alongside and a guy inside rolled down the window.

"Whores!" he screamed at them. The drive-by vitriol had nothing to do with the work, of course, but added a toxic splash to what was already a dreary experience. During their next meeting at Y Combinator, Garry Tan suggested that Dennis would be happier leaving. The team had only about six weeks to get ready for the Demo Day. Dennis agreed.

His departure solved nothing. The meme-generating project was inherently ridiculous, with vague ideas that groups of people would get together online to simultaneously pass around the remixed memes. Dan was miserable.

"Stuck in a house in the suburbs where the weather never changed, constantly around the same people, without a car," he said.

"Add the fact that we kicked Diaspora to the curb and were working on something that I thought was so dumb, I couldn't see a jail cell being any much worse."

Max soldiered on. Y Combinator was premised on relatively small bets spread around a bunch of start-ups. If only one from each batch succeeded, the funders would do well. Dropbox, for instance, had been a Y Combinator company. By the spring of 2012, companies that Y Combinator had nurtured during its first seven years were worth $7.8 billion in value. It had, to that point, no involvement with not-for-profit projects, no experience nurturing or sustaining them. (The fund supported its first nonprofit in January 2013, helping a site called Watsi, billed as a Kickstarter for medical care.)

Diaspora had no business being at Y Combinator because it had no business. And if it had prospects of a business—like the hosting services—it was hard to see them generating the explosive returns that drove the hedged bets being made. "Start-ups are an algorithm for solving problems," Max said. Though not the only algorithm, and not for all problems.

Makr.Io was a thin notion that would have been invisible had it not been attached to Diaspora. By August, it was clear that the team could not present Makr.Io at the Demo Day beauty pageant.

Meanwhile, the community of Diaspora coders and supporters was eroding from neglect. They had provided sustenance for Diaspora, the debugging and rewriting of code, the constant regeneration that keeps a piece of software alive. And they were an endlessly thoughtful, if contentious, group. But the attention of Max and Dan, the leaders of the project and the gatekeepers who approved proposed revisions, was absorbed by the Makr venture.

By the middle of August 2012, even the most devoted followers of Diaspora were falling off. One of them, Jason Robinson, who had been a regular donor to the project, posted a message noting that the demise of Diaspora was a perennial topic. "Personally before I have been a believer and have had faith in Diaspora Inc pulling off this whole thing. I have disregarded the doom talkers as people who just don't see the big picture. Today, I am one of those doom talkers.

"Currently my faith in this whole project is crumbling faster than ice in the polar region. There are so many things wrong that let's just list a few."

He noted that while they announced that they were working on Makr, they had said almost nothing about Diaspora, and were doing very little to maintain its code and the fixes that were being suggested.

"Guys I wish you could just reach out and say 'hey we're not interested any more, go home and do something else' if that is the case.

"Personally I am hoping some organization (Mozilla, etc) picks up and forks Diaspora* to build the next great social network that is modern and free. My trust in Diaspora Inc to be able to pull this off has dwindled—which I am very sorry to say since I think these guys have done an awesome job. The belief just doesn't seem to be there any more.

"As such, I am pulling my monthly donation to Diaspora Inc, as I no longer see that the products I am investing in will be something that is worth paying for. I am not paying to create Makr.io."

Robinson's well-reasoned critique was correct in every respect. He had always spoken up for the Diaspora team. Now he was supporting the project by speaking the plain truth. That others were harsher, or had slightly different takes on events, did not change the reality.

The summer was coming to an end, with no Makr to present at the demo session, and with Diaspora disintegrating. Max rarely slumped emotionally, but the convergence brought him down. One evening, they were invited to dinner by Danielle Morrill, the head of a Y Combinator group that was working on a social app called Referly, which had the aim of getting cash rewards for individuals who drove purchases of products that they wrote about. It was promising enough that Morrill did not have to worry about getting ready for Demo Day: she and her crew had spent most of the summer raising money, cooking big breakfasts, drinking a lot, and sleeping outside so they could watch meteor showers.

Over drinks in her yard, Morrill wanted to set Max straight. She wasn't buying the idea that Diaspora was a failure. What it had brought to the world would live on, she said, whether or not he and Rosanna and

Dan shifted their focus to other projects. It was normal to explore new interests. "Diaspora will not die," Morrill told them.

Her words rallied them. And there were hard numbers to back up what she was saying. No central body counted how many people were members of the Diaspora network—decentralization was, after all, the point of it—but a pod leader in Seattle, David Morley, had kept a running tally, an informal census. It showed that 125 pods were running, and that they had a total of 381,649 users. Max was damned if he was going to let Diaspora die. It might not have satisfied what they had envisioned, and then reenvisioned, but it was valuable to people around the world.

The next night, he huddled online with a young man in Peoria, Illinois, Sean Tilley, who had been trying to manage the Diaspora code forge for months while the team on the West Coast worked on Makr. They spent the weekend deciding how they were going to move forward.

That brought them to the chat room session on "Monday, Monday," August 27. It might have been ten A.M. in Sunnyvale, California, but hardly any of the other people online were in that time zone. Kevin Kleinman, the passionate young supporter, checked in from Holland, where it was seven P.M.

[12:47] <KevinKleinman> Good morning/evening everyone.

[12:50] <diegoch> Good morning/evening everyone !! Buenos días/tardes para todos !! Spanish users are also present :)

[12:51] <bla_> @diegoch german as well ;)

[12:51] <DeadSuperH[ero> Dang, 31 peeps in here. :)

That was only the start. Max put up one more video, tailored to his audience and the occasion: Elvis, singing "Suspicious Minds." More were checking in every minute, their handles popping onto screens across the world. Mathis. Inyan. ranmaruhibikiya. Makailija. Zeropoint. Goob. Otomo. alpHa OmeGa. ghosthand.

Max announced himself with a knockoff ghetto yell—on the Internet, no one knows that you're a dog, or a guy right out of college with a Land Rover parked in the driveway.

[12:58] <maxwell> holla!

[12:58] <DeadSuperHero> !

Code meetings had been held every other Thursday on the IRC channel, but they rarely were so well attended. The custom was that the meeting started four minutes late, to allow stragglers to keep up. As people logged in, it would take an instant and a half before their entries popped onto the screen, so there was a jagged asynchronous quality to the dialogue as one reply was being offered to a question that might have two or three others lodged after.

[13:04] <maxwell> whoop whoop

[13:04] <DeadSuperHero> Okay, it's that time!

The project had taken on a life beyond their original expectations, Max said.

[13:08] <maxwell> but right now, we think that we have to reflect this reality in the project itself

[13:08] <maxwell> so the purpose of this meeting today is that we are giving the code to the d* community who loves it so

[13:08] <shmerl> maxwell: Didn't you have far reaching plans to create FB alternative?

[13:08] <maxwell> so let this be an announcement that we are stepping forward and say, we are going to be creating some community governance

[13:09] <maxwell> so the thousands of people who love d* around the world can have a better say in what features are developed

[13:09] <goob> OK, how will this work?

[13:09] <Bugsbane> Let anarchy reign! \o/

[13:09] <maxwell> and what the priorities are

A governance structure had to be worked out, Max said. One partici-
pant then turned to the sticky question of the collective intellectual property built from hundreds of coders. From the day they threw open the
code foundry as part of the free-software development process, individual developers had been licensing their contributions to the body over-
seeing the project—in this case, Diaspora Inc., a for-profit corporation.
Max signaled that he and Dan, and any other shareholders who might
have a say, were ready to give it up.

[13:29] <maxwell> diasp: we need to step through this

[13:29] <maxwell> we want to put it in the right place

[13:29] <maxwell> so if we have to set up a holding org thats
community run

[13:29] <maxwell> we can do that

[13:30] <maxwell> i am open to stepping through the problem space
and figure out the right course of acton

[13:30] <shmerl> May be a diaspora foundation or something.

[13:30] <maxwell> yes

A foundation! It had been discussed for more than two years, and
now they were talking about it yet again. Nearly a year earlier, Yosem had
bought the domain name, DiasporaFoundation.org, that was intended to
be the website for the Diaspora foundation, which would accept donations of money and code. That set off the mutiny of October 2011. But
there was no foundation, and the subject simply had fallen off the radar.
Now it was being promised again.

Max and Sean made their good-byes and signed off. Rosanna had
weighed in a few times during the chat. Afterward, in a chat on Facebook
with me, she reflected that this step should have been taken much earlier:

"but i dont think there are really any regrets cause then we wouldnt
be where we are today

"its all part of the learning process

"just life

"things happen

"if life were that easy, everyone would have everything figured out

"you just have to work with whatever life gives you"

Dan was not heard from during the online meeting. Just as he had at Ilya's funeral, he remained silent. In fact, he slept through the session. He moved back to New York before the end of that week, and did not return to the West Coast. His father picked up his belongings.

Two days after Max's announcement, Gardner Bickford sat at sunrise in the Nevada desert with Ilya's bike. It was mounted nose down in the sand, precisely, intentionally. The bike Ilya rode all around San Francisco had been converted by some friends into a sculpture for the 2012 Burning Man gathering. Gardner, Mike Sofaer, and other friends had brainstormed and drew inspiration from the video of Ilya trying to drink a Bloody Mary on his bike one sunny weekend morning in the Mission.

Using light bars powered by a solar panel, they evoked the flight of a dandelion seed—evoking the logo for Diaspora—onto the ground, where his helmet rested.

They erected a wooden lectern, and set up a tablet computer with streaming photographs. For a while, there was a plan to serve Bloody Marys from the lectern, but the logistics of having an electrified sculpture in the middle of the playa were all-consuming, and they did not get around to making drinks. They did commission three hundred pairs of sunglasses stamped with the words "Begin with love."

At the end of the gathering, they dissembled the lectern and burned it. Gardner stored the bicycle in the RV that he kept in Fernley, Nevada, along with the dandelion effigy. He could picture that morning with Ilya sipping the Bloody Mary on his bicycle, preserved as it was on YouTube. It stayed in his memory as a single, revealing pixel.

"That was an event," Gardner said, "where he was attempting a feat, without any fear of failing."

# EPILOGUE

Finally back in New York, Dan got an apartment in Brooklyn and went to work for Pivotal's branch in Union Square, a short distance from NYU. He connected with a woman whom he had met in his first days of college, and they settled into a happy partnership.

Max and Rosanna set up a business in San Francisco, BackerKit, to help start-ups keep in touch with people who donated money in crowd-source fund-raising efforts. It was an idea that rose from Max's own experience of trying to ship T-shirts and stickers and also to make sure their six thousand Kickstarter supporters were kept up to date on the project—not to mention wishing there were a way to hit them up for additional funds. It was a sound idea: enough groups and people felt BackerKit's fee of 1 percent of the amount raised was worthwhile that the business became profitable within six months.

None of the three surviving founders of Diaspora was involved in the project on a daily basis. Sean Tilley and the leading contributors to the code were given control of the repository, deciding what proposed changes should be accepted. A guy in Spain rewrote all the software for the mobile site, keeping it up to date with new phones. Mr. ZYX, a young German, managed updates to translations. Someone else connected Foursquare, the location tracker and identifier. By March 2013, four major revisions of the code had been pushed out since the previous September.

Diaspora, the project, had refused to die. But Diaspora Inc., the for-profit corporation that the guys had set up in 2010, remained the holder of the copyright to all the thousands of lines of code. Who should own it? They needed a foundation, but creating one involved lawyers, accountants, IRS certification. Max ended up speaking with Eben Moglen, who had built the Free Software Support Network, which Moglen likened to a condominium for nonprofit organizations. Within it, free software projects could operate as subsidiaries, and the FSSN would take care of bookkeeping and taxes.

Moglen was interested, but wary.

"I don't want to be a project gravedigger," Moglen told Max. So what condition was Diaspora in?

In fact, hearty, under the control of the community. As Max would say: "They've pushed out four or five major updates. Honestly, they're running it a little more even-keeled than we were doing."

There was little sign of Max's involvement in the project, but behind the scenes, his discussions with Moglen's organization stretched on more than six months, with papers sent back and forth, various people not being available, and due diligence being conducted on both ends. Moglen's group had to be sure that Diaspora was viable; Max needed to understand how the structure would work.

In August 2013, Max made an announcement on the project's discussion board: Diaspora could join the FSSN if the community agreed. All the assets of Diaspora Inc., including the code and a few thousand dollars still in the bank, would go there. The volunteer community would run the project. Max suggested that they simultaneously launch a new crowd-sourced fund-raising effort, not because they had pressing needs for money but because he expected that these signs of vigor would get some favorable attention.

Max's sudden return to the online community with the proposition drew mixed responses—not on the merits but on the process. A major code contributor over the previous year, a European, Johann Haas, whose handle was Jonneβ, bristled that Max and Sean Tilley had been working on this transfer for months without consultation.

"Would it have been so hard to at least inform the community about the strategy you set before all the decisions had been made?" he asked on the forum.

Others, however, weighed in to defend Max and Sean, explaining that it was a proposal, not a final decision.

As a practical matter, if a group had been involved in assembling the granular details of the arrangement made by Max with Moglen's group, what had taken six months might well have turned into a death march. Instead of admiring the problem, Max had actually found a solution.

Still, there was no enthusiasm to do more fund-raising, as Max had suggested. Looking for support, he could not turn to Dan, who had withdrawn completely. He could, however, call on Rafi, who had graduated in May from NYU and spent the summer in New York as a teaching assistant. He was waiting to join the navy.

He supported Max publicly on the idea of turning the project over to the foundation, but shared the general reticence on the fund-raising. He, too, accused Max of presenting it as a fait accompli.

"A little bit guilty as charged," Max said.

More to the point, Rafi did not understand why funds had to be raised. No one was being paid. The expenses were very low. Max said that they didn't need to figure out where the money was going to go, but that they ought to strike when news broke of their resiliency, building up resources for things that would surely come along. Perhaps they could pay code bounties or organize gatherings. Or, for instance, they could sponsor fellows who would spend a few months working on nothing but the project.

Rafi was not persuaded. When Max went back online, Rafi was among those vocally opposed to raising money.

"We need to decide who will get the money," Rafi wrote in a public post.

"I love you buddy," Max replied, "but we had basically the same conversation three years ago, and if we acted then how you are suggesting then and now, we would have shut the Kickstarter down at 30k ☺. Surely we would not have been here three years later talking about it in front of a hundred people who love and care about our project."

The community overwhelmingly ratified the idea of accepting the assets of Diaspora Inc., and of setting up the project within the FSSN. The fund-raising proposal was turned down.

———

In late 2013 and into 2014, Edward Snowden, the former contractor for the United States National Security Agency, continued to make revelations about the American government's omnivorous appetite for surveillance across the world. In response, Tim Berners-Lee called for a global "Magna Carta" establishing the independence of the web and the rights of users on it. And with reports emerging that the NSA had impersonated Facebook users to plant spyware on their machines, Mark Zuckerberg spoke out in a blog post on his Facebook page. "I've called President Obama to express my frustration over the damage the government is creating for all of our future," Zuckerberg wrote.

Yet it hardly seemed possible to divorce the practices of governmental and commercial surveillance. In the fall of 2013, Eben Moglen gave four talks on "Snowden and the Future."

The government's ability to conduct wholesale global surveillance was made possible by "the great data-mining industry that has grown up in the United States, to surveil the world for profit, in the last fifteen years," Moglen said, comparing it with the industrial overreaching that was modifying the planet's climate. Analogously, the surveillance panopticon "is merely an ecological disaster threatening the survival of democracy."

The loss of privacy, Moglen said, was not the interception of a single kind of commerce or instance of communication between two people: "Privacy is an ecological rather than a transactional substance."

He cited two changes that had fundamentally altered the integrity of individual identity and privacy: e-mail and social networking: "They offer you free e-mail service, in response to which you let them read all the mail, and that's that. It's just a transaction between two parties. They offer you free web hosting for your social communications, in return for watching everybody look at everything."

On the notion that people had agreed to the monitoring by their use of the "free" services, he turned to the example of environmental law, which set standards for clean water and air. "You can't consent to expose your children to unclear or unsafe drinking water in the United States, no matter how much anybody pays you," he said.

All the huffing and puffing about the decrepit sense of modesty in young people—"humanity's overpublishing, that the real problem is the kids are sharing too darn much"—was misdirection, he said. If people were saying too much, then that could be a subject of complaint, but no more than that. "But really what has gone wrong is the destruction of the anonymity of reading, for which nobody has contracted at all," he said. "The anonymity of reading is the central, fundamental guarantor of freedom of the mind."

For the guidance of antiquity, Moglen—lawyer, hacker, and historian—turned to Edward Gibbon's volumes on the rise and fall of the Roman empire. From Scotland to Syria, the Romans had extended their empire by building roads that were still being used fifteen hundred years later. Down those roads, the emperor marched his armies.

"But up those roads he gathered his intelligence," Moglen said. "Augustus invented the posts: first for signals intelligence to move couriers and messages at the fastest possible speeds; and then for human intelligence."

With that infrastructure, Moglen said, "the emperor of the Romans made himself the best informed human being in the history of the world. That power eradicated human freedom."

There were easy ways to defeat the machinery of mass surveillance in the modern empire, Moglen said—or at least easy enough for those who worked with such technologies. The challenge was to spread it beyond the technical elite. "We must popularize it, make it simple, cheap, and easy—and we must help people put it everywhere," he said.

Or as the Diaspora guys had said: they had to make it for normal people.

Diaspora, inspired by Moglen's prophetic warnings more than three years before his Snowden lecture, had not penetrated popular culture; it was not a staple of world commerce; it was not both verb and noun, understood across the globe, like Facebook. Yet Diaspora, with its vision of decentralized, private communications centered on protecting the integrity of human connections, remains one of the most active open-source projects in the world. It has had more than fifty-five thousand lines of code, twenty-two revisions, hundreds of contributors over the course of its existence, and seventy-seven of them still regularly contributing code.

It had not died. It was not vaporware. The four NYU guys had not spent the money on appletinis and hookers. Its imaginative sparks, which could not be tracked completely, had lit up corners of the world far from the third-floor computer club at NYU. Diaspora had become even more necessary than when Dan, Rafi, Ilya, and Max had first been gripped by its possibilities.

As the project they created in 2010 moved to a foundation in 2013, Max wrote to the community that was going to be running it. "Diaspora is more than just code," he said. "It's about a glimmer of hope that a small group of people could actually make a profound change."

Or live trying.

# ACKNOWLEDGMENTS

Late one night in May 2010, Maxwell Salzberg, Daniel Grippi, and Raphael Sofaer spoke to me on a conference call for a newspaper column about their project. A few months later, I met Ilya Zhitomirskiy. By then, all four had agreed to let me chronicle their work for this book. From the beginning, they endured my presence and ignorance with nothing but generosity. They were idealists and gentlemen. It was a joy getting to know them.

Yosem Companys worked with them for eighteen months because he believed in their cause. From his records and extensive correspondence, I was able to grasp the turns that the project took, its aims and disappointments. In all our dealings, Yosem has been honest and diligent, a pleasure to learn from and to know.

Other people deeply involved in Diaspora, Rosanna Yao, Dennis Collinson, Sarah Mei, and Michael Sofaer, were generous in providing context for the events.

I profited immensely from Eben Moglen's lectures, as will anyone who searches for them online, and also from meeting with him a number of times during the research for this book.

Support from the Society for Professional Journalists through the Eugene C. Pulliam Fellowship enabled me, on the eve of the Arab Spring, to travel and meet journalists and activists of many countries. Among them were Walid Al-Saqaf, Houeïa Anouar, Mehdi Saharkhiz, Mohammad al-Abdallah, Chiarnuch Premchaiporn, Sharon Hom, Shivam Vij, and Esraa Rashid.

For insights and help, I thank Katitza Rodriguez of the Electronic Frontier Foundation, Jim Schuyler and Mike Rispoli, and Danny O'Brien, the journalist and civil libertarian.

I thank, among many who patiently tutored me in the nuts, bolts, and values of the open web, Tantek Celik and Blaine Cook. Also: Chris Messina, Evan Prodromou, Jon Phillips, Henrik Moltke, Mitchell Baker, Mark Surman, John Lilly, and Bill Woodcock. Harry Hamlin and Ian Jacobs, associated with the World Wide Web Consortium, gave me a primer on that organization's interests. My thanks to Tim Berners-Lee for speaking with me. Also, for inventing the world wide web.

Evan Korth and Biella Coleman, professors at New York University, who encouraged the explorations of the Diaspora Four and of many other students, helped me understand the spirit at the school when the four young men were students. Thanks also to Fred Benenson, Jamie Wilkinson, and Aditi Rajaram.

Finn Bruton, a postdoctorate fellow at NYU, first brought the Diaspora project to the attention of Cathy Dwyer, my wife, during a seminar run by Helen Nissenbaum, a leading privacy scholar. From Helen I learned about the tenth-century rabbi who prohibited opening other people's mail.

Among those in California who were generous with their time and insights, I thank Mitch Kapor, Randy Komisar, Janice Fraser, Rob Mee, Ian McFarland, and Aza Raskin.

I was helped by many friends of the four Diaspora principals, including Bobby Fishkin, Elizabeth Stark, Tony Lai, Gardner Bickford, David Kettler, Katie Johnson, Dan Goldenberg, Stephanie Lewkiewicz, Sashy Richmond, Adi Kamdar, Parker Phinney, and S. J. Klein. My thanks also to Carolyn Grippi, Casey Grippi, and Michael Wolf.

I had vital research help from Arikia Millikan, who spoke with many of Ilya's friends after his death, and from Lysandra Ohrstrom, who thoroughly briefed me on the ecology of apps and social networks.

At Viking, I thank Maggie Riggs and Georgia Bodnar, Randee Marullo, Noirin Lucas, John Thomas, Daniel Lagin, Gina S. Anderson, Nancy Sheppard, Carolyn Coleburn, Winnie de Moya and Louise Braverman.

While this book was being born, I had the support of editors and colleagues at *The New York Times*, including Susan Edgerley, Joe Sexton, Bill Schmidt, Phil Corbett, Carolyn Ryan, Wendell Jamieson, Gloria Bell, Noah Cohen, Jill Abramson, Dean Baquet, and Pete Khoury. The writing brothers Dolnick, Sam and Ben, passed along a great physical tool for writers (Scrivener) and my colleague and friend Emily S. Rueb shared an indispensable psychic contrivance (take it bird by bird, per Anne Lamott).

The enthusiasm and generosity of early readers, James McBride, Ken Auletta, Kevin Baker and Kevin Kelly, were most appreciated.

Through her dauntless representation of me over the last twenty-five years, Flip Brophy has made it possible for me to write books. My thanks, also, to her assistants, Holly Hillard and Julia Kardon.

Then there is the book's editor, Wendy Wolf, a relatively new presence in my life. Smart and painstaking and tactically brutal, she sees far and near, and is all you could ever wish for in an editor. Also, patient. And with bonus wisecracks. Other writers are envious when they hear about her, as well they should be.

On my way home from a reporting trip for this book, I stopped in Ireland and made my way to a farmhouse in Ower, County Galway, where the Molloy family has been making me welcome all my life.

Mary G. Murphy and T. J. Reagan fed me and let me hang out with them in San Francisco. Making friends with Mary in high school more than forty years ago turns out to have been a pretty good idea, so far.

Phil Dwyer, my dad and neighbor, listened to me talk about this book during breakfast at Vicky's diner and just as he always had, did everything he could to make it possible for me to get to the work I wanted.

In concrete ways, Cathy, Maura, and Catherine, wife and daughters, opened my eyes to the power of social media. Maura arrived for freshman year at college in 2004, just as a brand new thing called Facebook was being introduced. She and her roommate soon found that they were checking in so obsessively that during exams, they had to change each other's passwords to avoid temptation. Cathy noticed; she herself was starting work toward her doctorate, and wound up writing her dissertation on privacy management in Facebook and MySpace. At the time, hardly anyone was paying attention to the subject.

It was Cathy who told me about Diaspora in 2010 and why it mattered. She was right and wise.

If you are among the many people who helped me with this work but are not mentioned here, please mark the lapse as a failure of wit and order on my part. It is not a shortage of gratitude. Thank you.

# AUTHOR'S NOTE

**M**ore *Awesome Than Money* follows the evolution of the Diaspora
project between September 2009 and September 2013. I started
paying attention in May 2010, first to write a newspaper column
about it and then to chronicle its development. I did that primarily by
spending time with the four cofounders in California and New York as
they worked, played, dreamed, and ate. I witnessed many of the events
described here or was in the room for a lot of the conversations repro-
duced here. But many things I did not see or hear myself. Instead, I was
told about them in detail from people who were present. Where important differences in recollections among sources could not be recon-
ciled, the alternative versions are presented. In virtually every case, this
account rests on two or more on-the-record sources, including interviews
with the four founders, their friends, teachers, and associates, e-mail
exchanges, online discussions, and other documentary sources such as
contemporaneous notes or Youtube video. The names of the interviewees
can be found in the acknowledgments, beginning on page 347. No
unnamed sources were used in the making of this book. None of the subjects sought or was offered editorial control of the contents.

That books are static is intrinsic to their value. Fixing perspective at
a given moment makes understanding events and their context a manageable undertaking. This book covers what happened, as best I could

figure it out, during four years when dictators were overthrown, revolutions were launched and turned back, companies merged, and whistles blown. No less than our own skin, the membranes of technology and society are dynamic: the configurations of code, law, and businesses evolve by the instant. You never step into the same browser twice.

# NOTES

**EPIGRAPH**

vii **There's something deeper:** Emily Nussbaum, "Defacebook," *New York Magazine*, September 26, 2010.

**INTRODUCTION**

8 **"Facebook holds and controls":** Eben Moglen, prepared testimony for Congress, December 2, 2010, http://www.softwarefreedom.org/events/2010/do-not-track/Eben-Moglen-2010-12-2-privacy-testimony.pdf.

8 **At the headquarters:** James Harkin, "The Facebook Effect by David Kirkpatrick," *The Guardian*, July 17, 2010, http://www.theguardian.com/technology/2010/jul/18/the-facebook-effect-david-kirkpatrick-book-review.

8 **Atlas Solutions, purchased:** Eliza Kern, "Facebook Purchases Microsoft's Atlas Solutions for Reported $100 Million," *Gigaom*, February 28, 2013, http://gigaom.com/2013/02/28/facebook-purchases-microsofts-atlas-solutions-for-reported-100-million/.

8 **And virtually unknown to users:** Edmund Sanders, "AOL Cookies, Web Bugs, to Track Advertising," *Los Angeles Times*, October 5, 2001, http://articles.latimes.com/2001/oct/05/business/fi-53680.

9 **"What They Know":** A comprehensive series on tracking by the *Wall Street Journal*, accessible online at http://online.wsj.com/public/page/what-they-know-digital-privacy.html.

10 **In mid-2013:** Savik Das, Adam Kramer, "Self Censorship on Facebook," a paper presented on July 8, 2013, at the International Conference on Weblogs and Social Media, and posted online at http://sauvik.me/system/papers/pdfs/000/000/004/original/self-censorship_on_facebook_cameraready.pdf?1369713003.

10 **That means:** Jennifer Golbeck, "On Second Thought: Facebook wants to know why you didn't publish that status update you started writing," December 13, 2013, *Slate*. http://www.slate.com/articles/technology/future_tense/2013/12/face book_self_censorship_what_happens_to_the_posts_you_don_t_publish.html.

10 **Almost everyone on Planet Earth:** Eben Moglen, "The Union May It Be Preserved," a talk given at Columbia University School of Law, November 13, 2013, http://snowdenandthefuture.info/PartIII.html.

10 **One day in the 1970s:** Author's interview with Doug Engelbart, April 9, 1997.

11 **A little-celebrated figure:** John Markoff, "Computer Visionary Who Invented the Mouse," *New York Times,* July 3, 2013, http://www.nytimes .com/2013/07/04/technology/douglas-c-engelbart-inventor-of-the-computer-mouse-dies-at-88.html?pagewanted=all.

11 **The cofounder of Apple:** Jessica Guynn, "Douglas Engelbart Dies at Age 88; Computer Visionary," *Los Angeles Times,* July 3, 2013, http://www.latimes .com/news/obituaries/la-me-douglas-engelbart-20130704,0,1651153.story.

11 **Writing in the *Atlantic*:** Alexis C. Madrigal, "The Hut Where the Internet Began," *Atlantic,* July 7, 2013, http://www.theatlantic.com/technology /archive/13/07/the-hut-where-the-internet-began/277551/.

12 **By 2010, in just the two years:** Eric Eldon, "New Facebook Statistics Show Big Increase in Content Sharing, Local Business Pages," *Inside Facebook,* February 15, 2010, http://www.insidefacebook.com/2010/02/15/new-facebook-statistics-show-big-increase-in-content-sharing-local-business-pages/.

12 **A manager in Facebook's growth:** Andy Johns, "What Are Some Decisions Taken by the 'Growth Team' at Facebook that Helped Facebook Reach 500 million users?" *Curaqion,* October 2012, http://curaqion.com/issue/01/facebook-growth.

## CHAPTER ONE

18 **a commonplace computer language:** http://www.facebook.com/blog/blog .php?post=2356432130.

18 **10 million websites:** http://www.php.net/usage.php.

29 **two researchers at the University of Texas:** http://www.cs.utexas.edu /~shmat/shmat_oak08netflix.pdf.

30 **The digital bread crumbs led:** http://www.nytimes.com/2006/08/09 /technology/09aol.html?_r=2&oref=login&pagewanted=all&.

## CHAPTER TWO

32 **a piece of hardware:** From Max, August 8, 2010.

35 **Ilya was rewarded:** Talk by Robert Dorf at Ilya's memorial service, November 20, 2011.

41 **In 2013, Makerbot:** Ashlee Vance and Joshua Brustein, "MakerBot Sells Out to 3D Printing's Old Guard for $403 Million," *Bloomberg Businessweek,* June 19, 2013, http://www.businessweek.com/articles/2013-06-19/makerbot-sells-to-3d-printings-old-guard-for-403m.

41 **The work of Turing:** George Dyson, "An Artificially Created Universe": The Electronic Computer Project at IAS," *The Institute Letter,* The Institute for Advanced Study, Spring 2012, http://www.ias.edu/about/publications/ias-letter /articles/2012-spring/george-dyson-ecp.

## CHAPTER THREE

55 **They were so obviously struggling:** Sarah Perez, "Diaspora Project: Building the Anti-Facebook," May 5, 2010, *ReadWriteWeb,* www.readwriteweb .com/archives/diaspora_project_building_the_anti-facebook.php.

## CHAPTER FOUR

61 **Outside, the reaction was brutal:** Ryan Singel, "Facebook's Gone Rogue; It's Time for an Open Alternative," *Wired,* May 7, 2010. http://www.wired .com/business/2010/05/facebook-rogue/.

62 **Around that time:** Michelle Madejski, Maritza L. Johnson, and Steven M. Bellovin. "A Study of Privacy Settings Errors in an Online Social Network," PerCom Workshops, pp. 340–45. IEEE (2012), https://www.cs.columbia.edu/ ~smb/papers/fb-violations-sesoc.pdf.

67 **They were dizzy:** Komisar interview with me and his e-mail to Yosem:

> From: Randy Komisar <rkomisar@kpcb.com>
> Date: Sat, May 8, 2010 at 2:30 PM
> To: Yosem Companys <companys@stanford.edu>
> Thx
> Friday we all agreed they would come out for the summer and work with us in the incubator.
> Best
> rau

69 **At midnight, my column was posted:** Jim Dwyer, "Four Nerds and a Cry to Arms Against Facebook," *New York Times,* May 11, 2010, http://www.nytimes .com/2010/05/12/nyregion/12about.html?_r=0.

## CHAPTER FIVE

77 **In an interview with *Wired:*** Ryan Singel, "Mark Zuckerberg: I Donated to Open-source Facebook Competitor," *Wired,* May 28, 2010, http://www.wired .com/business/2010/05/zuckerberg-interview/all/.

77 **"The funding is not through the capital":** Ryan Singel, "Open Facebook Alternatives Gain Momentum, $115K," *Wired,* May 13, 2010, http://www .wired.com/business/2010/05/facebook-open-alternative/.

## CHAPTER SIX

78 **Democratic revolutions followed its adoption:** Bertrand de La Chapelle, a deputy in the French ministry of Foreign Affairs, remarks at "Internet at

Freedom" conference, Central European University, Budapest, September 2010.

79 **A crusading blogger:** Cynthia Johnston, "YouTube Stops Account of Egypt Anti-Torture Activist," *Reuters,* November 27, 2007, http://www.reuters .com/article/2007/11/27/egypt-youtube-idUSL2759043020071127.

81 **Censorship was becoming a global norm:** Peter Eckersley, "2010 Trend Watch Update: Global Internet Censorship," Electronic Frontier Foundation, December 10, 2010, https://www.eff.org/deeplinks/2010/12/2010-trend-watch-update-global-internet-censorship.

81 **The filtering software was provided:** Websense discusses the situation in Yemen and says it will stop updating the subscriptions there, http: //community.websense.com/blogs/websense-features/archive/2009/08/17 /websense-issues-statement-on-use-of-its-url-filtering-technology-by-isps-in-yemen.aspx.

81 **"Often pitched in the first instance":** Helmi Norman and Jillian C. York, "West Censoring East: The Use of Western Technologies by Middle East Censors, 2010–2011," OpenNet Initiative, March 2011, http://opennet.net /west-censoring-east-the-use-western-technologies-middle-east-censors-2010-2011.

82 **The strongest hands played:** James Ball, Bruce Schneier, and Glenn Greenwald, "NSA and GCHQ Target Tor Network That Protects Anonymity of Users," *Guardian,* October 4, 2013, http://www.theguardian.com/world/2013/oct/04 /nsa-gchq-attack-tor-network-encryption.

83 **Google released its first "transparency report:** "Government Requests for Information Double over Three Years," November 14, 2013, *Google Official Blog,* http://googleblog.blogspot.com/2013/11/government-requests-for-user.html.

84 **The agents had a powerful tool:** "Making the world safer with trend-setting intelligence solutions," Nokia manual, published online at http://www.voima.fi /tiedostot/NSN_Image_Bro_web1.pdf.

84 **The extent of this power:** Hanna Nikkanen, "Technology Failed Iran," *Voima,* March 1, 2010, http://fifi.voima.fi/artikkeli/Technology-failed-Iran /3407?page=1.

84 **European regulators had themselves:** Ibid.

84 **"Governments in almost all nations required operators":** Statement from Barry French, executive board member and head of marketing and corporate affairs, Nokia Siemens, "Networks, Hearing on New Information Technologies and Human Rights," European Parliament, Subcommittee on Human Rights, Wednesday, June 2, 2010.

85 **An American company called:** John Markoff, "Rights Group Reports on Abuses of Surveillance and Censorship Technology," *New York Times,* January 16, 2013, http://www.nytimes.com/2013/01/16/business/rights-group-reports-

on-abuses-of-surveillance-and-censorship-technology.html?ref=technol-ogy&_r=0.

## CHAPTER SEVEN

90 **"Consider that informed citizens":** Richard Esguerra, "An Introduction to the Federated Social Web," Electronic Frontier Foundation, March 21, 2011, https://www.eff.org/deeplinks/2011/03/introduction-distributed-social-network.

## CHAPTER EIGHT

94 **It billed the services:** Amount quoted to Max, and confirmed as approximately right by Mike Sofaer.

97 **It came to a close:** Interviews with Dan, Max, Ilya, and Mike himself.

102 **Bill Gore, the inventor of Gore-Tex fabric:** NPR staff, "Don't Believe Facebook; You Only Have 150 Friends," *NPR*, June 5, 2011, http://www.npr.org/2011/06/04/136723316/dont-believe-facebook-you-only-have-150-friends.

102 **Whether they were picking bugs:** Guillaume Dezecache, "Are We Sure We Can Groom Beyond Dunbar's Number?" June 4, 2012, International Cognition and Culture Institute, http://www.cognitionandculture.net/workshops/77-dunbars-number/2415-are-we-sure-we-can-groom-beyond-dunbars-number.

110 **"People knowledgeable about Google":** Nicole Perlroth, John Markoff, "N.S.A. May Have Hit Internet Companies at a Weak Spot," *New York Times,* November 25, 2013, http://www.nytimes.com/2013/11/26/technology/a-peephole-for-the-nsa.html?ref=johnmarkoff&_r=0.

111 **"modern cars are computers":** Cory Doctorow, "The Coming Civil War over General Purpose Computing," a talk delivered at Google in August 2012, http://boingboing.net/2012/08/23/civilwar.html.

## CHAPTER NINE

114 **"This message has been brought to you":** https://blog.mozilla.org/press/files/2013/11/nytimes-firefox-final.pdf.

115 **By October 1994:** http://www.wired.com/wired/archive/2.10/mosaic.html.

118 **A developer named Jamie:** http://www.jwz.org/gruntle/nscpdorm.html.

118 **Four months later:** http://news.morningstar.com/articlenet/article.aspx?id=741.

119 **The level of vitriol:** http://www.wired.com/wired/archive/2.10/godwin.if_pr.html.

121 **"There was no interest":** http://www.wired.com/business/2012/05/epicenter_isocfamersqabaker/.

122 **In a series of communiqués:** Vulnerability Note VU#413886, Vulnerability Notes Database, February 2, 2004, updated October 28, 2004, http://www.kb.cert.org/vuls/id/413886, October 13, 2004.

122 **The Mozilla staff of fourteen:** Steve Lohr and John Markoff, "In the Battle of the Browsers '04, Firefox Aims at Microsoft," *New York Times,* November 15, 2004, http://www.nytimes.com/2004/11/15/technology/15browser.html?ex=1258261200&en=012c4675cd53ba55&ei=5090&partner=rs&_r=1&.

123 **Then an e-mail arrived:** Gervase Markham, "Parties on Seven Continents," *Hacking for Christ,* October 21, 2004, http://blog.gerv.net/2004/10/.

124 **The privacy setting became known:** Gaurav Aggarwal, Elie Bursztein, Collin Jackson, and Dan Bone, "An Analysis of Private Browsing Modes in Modern Browsers," presented at Usenix 2010 Conference, Washington, D.C., http://crypto.stanford.edu/~dabo/pubs/papers/privatebrowsing.pdf.

125 **By 2012, Firefox's annual:** http://www.zdnet.com/blog/networking/firefox-hits-the-jackpot-with-almost-billion-dollar-google-deal/1780.

125 **"Some people say, 'I don't care'":** Interview with Mitchell Baker, October 6, 2010.

126 **Aza Raskin, the twenty-six-year-old:** Interview with Raskin, June 2013.

## CHAPTER TEN

130 **The 2010 theme**: Video, http://vimeo.com/14733288.

135 **"Jonah Peretti":** http://www.shey.net/niked.html. Timothy Shey, "The Life of an Internet Meme," published on Shey's weblog, tim.shey.net, http://www.shey.net/niked.html.

136 **He also appeared:** Ibid.

## CHAPTER ELEVEN

140 **Heap said he would build:** Charles Arthur, "Haystack Anticensorship Software Withdrawn Over Security Concerns," *The Guardian,* September 17, 2010, http://www.theguardian.com/technology/2010/sep/17/haystack-software-security-concerns.

140 **It received vital licenses:** Indira Lakshmanan, "Interview with Hillary Rodham Clinton," March 19, 2010, Bloomberg TV, http://www.state.gov/secretary/rm/2010/03/138677.htm.

140 **At the moment a totalitarian:** Evgeny Morozov, "The Great Internet Freedom Fraud," *Slate,* September 16, 2010, http://www.slate.com/articles/technology/technology/2010/09/the_great_internet_freedom_fraud.html.

141 **Meanwhile, the acclaim mounted:** Aleks Krotoski, "Media Guardian Innovation Awards," *Guardian,* March 28, 2010, http://www.theguardian.com/media/2010/mar/29/austin-heap-megas-innovator-award.

147 **Dozens of types of free licenses:** "Various Licenses and Comments About Them," GNU Operating System website, https://www.gnu.org/licenses/license-list.html.

148 **McKenzie was interviewed:** Dan Goodin, "Code for Open Source Facebook Littered with Landmines," *Register,* September 16, 2010, http://www.theregister.co.uk/2010/09/16/diaspora_pre_alpha_landmines/.

149 **Writing in a blog later:** Patrick McKenzie, "Security Lessons from the Diaspora Launch," *Kalzumeus,* http://www.kalzumeus.com/2010/09/22/security-lessons-learned-from-the-diaspora-launch/.

## CHAPTER THIRTEEN

171 **"What if you build a social network":** Christina Warren, "Hands-on with Facebook Alternative Diaspora," *Mashable,* November 24, 2010, http://mashable.com/2010/11/24/diaspora-preview/.

171 **On *ZDNet,* another important tech:** Dana Blankenhorn, "Is Diaspora Too Late?" *ZDNet,* November 24, 2010, http://www.zdnet.com/blog/open-source/is-diaspora-too-late/7877.

171 **And on *Ars Technica:*** Ryan Paul, "Hands On: A First Look at Diaspora's Private Alpha," *Ars Technica,* November 29, 2010, http://arstechnica.com/information-technology/2010/11/hands-on-a-first-look-at-diasporas-private-alpha-test/.

172 **Sarah Mei, the developer:** Sarah Mei, "Disalienation: Why Gender Is a Text Field on Diaspora," November 26, 2010, SarahMei.com, http://www.sarahmei.com/blog/2010/11/26/disalienation/.

173 **"Facebook develops new features":** Patricio Robles, "Diaspora's Gender Field Controversy and the Consumer Internet," *Econsultancy,* November 30, 2010, http://econsultancy.com/us/blog/6909-diaspora-s-gender-field-controversy-and-the-consumer-internet.

173 **Perhaps the most striking reaction:** Tim Berners-Lee, "Long Live the Web: A Call for Continued Open Standards and Neutrality," *Scientific American,* November 22, 2010.

## CHAPTER FOURTEEN

175 **Almost immediately, a massive cyberattack:** Ryan Paul, "Wikileaks Moves to Amazon's Cloud to Evade Massive DDoS," *Ars Technica,* November 30, 2010, http://arstechnica.com/security/2010/11/wikileaks-moves-to-amazons-cloud-to-evade-massive-ddos/.

175 **Within a day, Amazon:** Ryan Paul, "Wikileaks Kicked out of Amazon's Cloud," *Ars Technica,* December 1, 2010, http://arstechnica.com/security/2010/12/wikileaks-kicked-out-of-amazons-cloud/.

176 **"Despite its name, 'cyberspace'":** Nate Anderson, "Where's Wikileaks? The 'infowar' is on as site hops servers," *Ars Technica,* December 3, 2010, http://arstechnica.com/tech-policy/2010/12/wheres-wikileaks-the-infowar-is-on-as-site-hops-servers/.

179 **There were many others, including Ilya:** account of Ilya's participation in Liberation Technology activities during the Arab Spring provided by Yosem Companys in e-mail March 19, 2013, and in interviews March–April 2013.

## CHAPTER SIXTEEN

199 **Instead, sipping the Bloody Mary:** Video of Ilya's bike ride with the Bloody Mary, http://www.youtube.com/watch?v=iY2ilf7KSqs.

200 **The class was called Ideas:** Anya Kamenetz, "The Most Influential Women in Technology 2010—Elizabeth Stark," *Fast Company,* March 24, 2010, http://www.fastcompany.com/1596380/elizabeth—cofounder-open-video-alliance.

207 **"they were looking for office space":** Pascal Finette, e-mail exchange with Jim Dwyer, March 8, 2013.

## CHAPTER EIGHTEEN

214 **"Ilya came up":** E-mail read to me by Adi on April 11, 2013, in phone interview.

216 **At the beginning:** Natasha Singer, "A Vault for Taking Charge of Your Online Life," *New York Times,* December 8, 2012, http://www.nytimes.com/2012/12/09/business/company-envisions-vaults-for-personal-data.html?pagewanted=1&_r=1&hp&adxnnlx=1354996826-eujerCBnyNUsOGDSyaoexA&.

218 **One tech blog declared:** ITProPortal staff writer, "Hands On with Google+," July 4, 2011, http://www.itproportal.com/2011/07/04/hands-on-google-plus/#ixzz2N5ChtsUa.

218 **John Henshaw of the *Raven:*** "Google+ Runs Circles Around Diaspora," *Raven,* June 29, 2011, http://raventools.com/blog/google-copies-diaspora/.

218 **That was why it had:** Google Form 10K for fiscal year ending December 31, 2012, http://edgar.secdatabase.com/1404/119312513028362/filing-main.htm.

## CHAPTER NINETEEN

222 **The guys were giving him flak:** E-mail from Max to Yosem, July 6, 2011.

224 **Google, in fact, had kept:** "Why Is Almost Half of Google in Beta?" *Royal Pingdom,* September 24, 2008, http://royal.pingdom.com/2008/09/24/why-is-almost-half-of-google-in-beta/.

225 **More than 50 million blogs:** Matt Brian, "WordPress: Now Powering 50 Million Blogs," *The Next Web,* July 10, 2011, http://thenextweb.com/media/2011/07/10/wordpress-now-powering-50-million-blogs/.

## CHAPTER TWENTY

232 **Within two months of its creation:** Lauren Indvik, "Facebook: Zynga Generates 12% of Our Revenues and We Need Them," *Mashable,* February 1, 2011, http://mashable.com/2012/02/01/zynga-facebook-revenue/.

234 **The lowest price:** Dylan (one name), "Burning Man Festival 2012: Preview," *SeatGeek,* July 26, 2011, http://seatgeek.com/blog/concerts/burning-man-festival-2012-tickets.

## CHAPTER TWENTY-ONE

243 **In July, Randi Zuckerberg:** Bianca Bosker, "Facebook's Randi Zuckerberg: Anonymity Online 'Has to Go Away,'" *Huffington Post,* July 27, 2011, http:

//www.huffingtonpost.com/2011/07/27/randi-zuckerberg-anonymity-
online_n_910892.html.

243 **Eric Schmidt, the former CEO:** Bianca Bosker, "Eric Schmidt on Privacy:
Google CEO Says Anonymity Online Is 'Dangerous,'" *Huffington Post,* August
10, 2010, http://www.huffingtonpost.com/2010/08/10/eric-schmidt-privacy-
stan_n_677224.html.

## CHAPTER TWENTY-TWO

254 **To rebut him:** Martin Kaste, "Who Are You, Really? Activists Fight for Pseudonyms," *NPR,* September 28, 2011, http://m.npr.org/news/front/140879480.

## CHAPTER TWENTY-THREE

262 **How often did start-ups fail:** Deborah Gage, "The Venture Capital Secret: 3
out of 4 Startups Fail," *Wall Street Journal,* September 20, 2012, http://online
.wsj.com/news/articles/SB10000872396390044372020457800498047642919O.

262 **As Drew Houston, who started:** Drew Houston, "MIT Commencement
Address," *MIT News,* June 7, 2013, http://web.mit.edu/newsoffice/2013/
commencement-address-houston-0607.html.

## CHAPTER TWENTY-SIX

287 **Officials in San Francisco:** Kevin Fagan, "Occupy SF: No Show Raid Invigo-
rates Protesters," *San Francisco Chronicle,* October 27, 2011, http://www
.sfgate.com/default/article/Occupy-SF-No-show-raid-invigorates-
protesters-2325517.php.

## CHAPTER TWENTY-EIGHT

298 **Ilya had been blown away:** John Brockman, "The Local Global Flip, or the
Lanier Effect," *Edge,* August 29, 2011, http://www.edge.org/conversation
/the-local-global-flip.

300 **technology news section of the *Wall Street Journal:*** Nick Clayton, "Whatever Happened to Diaspora, the 'Facebook Killer'?" *Wall Street Journal,*
November 7, 2011, http://blogs.wsj.com/tech-europe/2011/11/07/whatever-
happened-to-diaspora-the-facebook-killer/.

## CHAPTER THIRTY-TWO

332 **Nearly two years to the day:** Karen Weise, "On Diaspora's Social Network,
You Own Your Data," *Bloomberg Businessweek,* May 10, 2012, http://www
.businessweek.com/printer/articles/24762-on-diasporas-social-network-
you-own-your-data.

335 **By the spring of 2012:** Tomio Geron, "Top Startup Incubators and Accel-
erators: Y Combinator Tops with $7.8 Billion in Value," *Forbes,* April 30, 2012,
http://www.forbes.com/sites/tomiogeron/2012/04/30/top-tech-incubators
-as-ranked-by-forbes-y-combinator-tops-with-7-billion-in-value/.

336  **she and her crew had spent:** Danielle Morrill, "Reflecting on My Career at the 10 Year Mark," *Referly,* http://refer.ly/reflecting-on-my-career-at-the-10-year-mark/c/855dd3b2765a11e2bfbf22000a1db8fa.

**EPILOGUE**

344  **Tim Berners-Lee called for:** Jemima Kiss, "An Online Magna Carta: Berners-Lee Calls for Bill of Rights for Web," *The Guardian,* March 11, 2014, http://www.theguardian.com/technology/2014/mar/12/online-magna-carta-berners-lee-web.

344  **And with reports emerging:** Ryan Gallagher and Glenn Greenwald, "How The NSA Plans to Infect 'Millions' of Computers with Malware," *The Intercept,* March 12, 2014, https://firstlook.org/theintercept/article/2014/03/12/nsa-plans-infect-millions-computers-malware/.

344  **Mark Zuckerberg spoke out:** Mark Zuckerberg, Facebook post,  March 13, 2014, https://www.facebook.com/zuck/posts/10101301165605491?stream_ref=1.

# INDEX